Ecology and Human Adaptation

ELEMENTS OF ANTHROPOLOGY
A Series of Introductions

Ecology and Human Adaptation

William A. Stini
University of Kansas

WM. C. BROWN COMPANY PUBLISHERS
Dubuque, Iowa

ANTHROPOLOGY SERIES

Library of Congress Catalog Card Number: 74-17708

ISBN 0—697—07537—0

Printed in the United States of America

To Ruth

Contents

	Preface	ix
1	What Is Adaptation?	1
2	The Species Homo Sapiens in Taxonomic Perspective	14
3	Genetic Adaptation: Evidence of Darwinian Evolution in Human Populations	34
4	The Human Capacity to Adjust	52
5	How To Succeed in Adaptation Without Really Trying	71
	Glossary	75
	Index	79

Preface

Physical anthropology has undergone a number of significant changes during the last twenty-five years. Interest in human evolution still retains the central position in the theoretical framework of the discipline, but increased emphasis on mechanisms has generated a demand for greater understanding of many areas of biological sciences. Man's origins and the meaning of the fossil evidence of our ancestry are still areas of concern to physical anthropologists. But interest in the course of ongoing human evolution has been added and now concerns a large proportion of those who study the biology of man.

As a result of these additional interests there has been a proliferation of subdisciplines, each relying upon the mastery of specialized techniques to collect data bearing on some aspect of human biology. So it is that biological anthropology is a discipline professed by researchers whose methods are those of geneticists, physiologists, anatomists, biochemists, biophysicists, and a number of other biological sciences. It is now possible to develop a complete coverage of general biology within the framework of human biology as a result of these developments in the field.

An important factor leading to increasing recognition that biological anthropology has much to contribute to the field of general biology as well as to anthropology has been a widening interest in ecology. Along with the realization that human intervention has already made significant changes in our environment and resultant fears of what the future will bring has been an increased recognition that our species is one of many living on this planet. Recognition of the interdependency of living things leads to the conclusion that damage to other species may, in yet unknown ways, bear serious consequences to our own survival.

Placing the study of human biology in this frame of reference we are led to inquire about the nature of the human ecological niche. Even if it is not true that "man is the master of all he surveys," it would be wise for us to know exactly where we fit in the overall scheme of life on this planet. What factors led to the attainment of so much control over the forces of nature and the resources of the planet by a single, none too impressive, mammalian species? This is, of course, a historical question and in many ways, the study of evolution is historical.

The focus of evolutionary change is the

process of adaptation, a process which generates new capabilities and creates new demands on the organisms undergoing it. The understanding of this process pulls all of the biological sciences together in the effort to retrace our origins and formulate a prognosis for our future. This is the concern of this book.

The interests of biological anthropologists have become more involved in the biology of living humans and their relationships to their environment. In this endeavor we are increasingly dependent upon our colleagues among cultural anthropologists for insights concerning the manner in which human groups alter the adaptive status of their members. Much of contemporary human adaptation is involved in being a member of a culture. The shape of future human evolution cannot help but be shaped by the activities of human society. So it is essential that the dialogue between cultural and biological anthropology focusing on a mutual concern for ecology and human adaptability be encouraged.

1 What Is Adaptation?

The word "adaptation" has been used in many ways. It is at the same time both highly descriptive and grossly misleading. Used in the vernacular sense, the most acceptable definition would imply the occurrence of change, making the organism or population more suitable for the environment in which it dwells. Since the environment is itself capable of change, organisms are continually confronted with new problems, problems which must evoke a change in the adapting species if survival in a competitive world is to be possible.

When biologists attempt to define "adaptation" more rigorously, serious problems arise. This is because biological adaptations occur in a variety of ways. Some adaptations are of a "permanent" nature, i.e., represent changes in the genetic constitution of the population. This type of adaptation is the only one which many biologists find truly deserving of the designation "adaptation."[1] To scientists favoring this point of view, among whom are many geneticists, no adaptation has occurred if the population has not experienced some alteration in gene frequencies. Adjustments which can occur within the lifetime of an individual and are not acquired specifically through inheritance or transmitted to succeeding generations fail to qualify as adaptations and must therefore be differentiated from true adaptations. There is considerable merit in this point of view. It is, in any event, clear and relatively unambiguous. By excluding adjustments that are not directly reflective of an altered genetic constitution, a rigidly genetic definition of adaptation places the term in a clearly evolutionary perspective. In fact, pursued to its logical conclusion, this choice of definitions allows us to say, along with Wallace and Srb, "Adaptation is evolution."[2]

However, in order to understand the process of evolution, by which any species adjusts to its environment, it is necessary to consider some of the ways overspecialization can be avoided. Avoidance of overspecialization is also, as will be shown in later sections of this book, an important adaptive strategy. Since humans have be-

1. For discussion see: Jack T. Stern, Jr., "The Meaning of 'Adaptation' in Relation to the Phenomenon of Natural Selection" in *Evolutionary Biology*, Vol. 4, eds. T. Dobzhansky, M. K. Hecht and W. C. Steere (New York: Appleton-Century-Crofts, 1970).

2. Bruce Wallace and Adrian Srb, *Adaptation* (Englewood Cliffs, N.J.: Prentice-Hall, 1964), p. 9.

come the species most adept at adjusting to a wide variety of environments through nongenetic mechanisms, we will find it essential to consider a number of categories of "adaptation" which do not satisfy the more rigorous definition of the word preferred by many geneticists. When nongenetic adaptations become the object of scrutiny, one of the first distinctions to emerge is the one which recognizes the difference between the "genotype" and the "phenotype." It is here that we can first get some notion of why "adaptation" has become a term to use with caution.

GENOTYPE AND PHENOTYPE IN EVOLUTIONARY PERSPECTIVE

Even when a given trait is under genetic control, it may express itself to varying degrees in response to the interaction between the genetic program inherited by an individual and the environmental pressures encountered, usually early in life. In order to express the distinction between the genetic potential inherited from the parents and the characteristics of the individual which actually can be seen, identified, and measured, the terms "genotype" and "phenotype" have been adopted. The genotype includes all of the genetic potential, whether expressed or unexpressed, inherent in the DNA sequences obtained by the fertilized egg (zygote) from the gametes (sperm and egg) which were contributed by the parents. It is not known how many "genes" (or genetic loci) collaborate to make up the genotype of a human being, but the estimates range upward from 20,000 loci, or gene pairs per individual. Since these loci are reassorted independently of each other when gametes are formed, it is clear that the number of gene combinations or genotypes possible in our species is very large if we assume that there is some variability possible at a number of these loci. Among the significant findings of much current genetic research has been that a surprisingly large number of loci does indeed possess such variability (allelic variation).[3] In many cases, it has been found that many alleles occur at a single locus. The net result of the occurrence of many genetic variations at a number of loci, independently reassorting in the process of sexual reproduction, is the provision of a vast number of potential genotypes to be tested against the environment. If we combine the observation of these mechanisms for the production of genotypic variability, mechanisms strongly dependent upon the process of sexual reproduction, with the additional observation that sexual reproduction is extremely widespread among living forms, we may reasonably infer that the maintenance of variability enhances the survivorship of species. In addition, sexual reproduction, by allowing different lines of descent to exchange newly arisen variability produced by mutation, facilitates the spread of beneficial new traits in a population.

Some alleles are dominant over their partners and thereby prevent their expression in the heterozygote. As a result, some genetic traits which are potentially harmful when expressed are carried from generation to generation without being lost to natural selection. Most mutations are less beneficial than their normal alleles (which, after all, are "normal" because they have passed the test of natural selection over many generations). Therefore,

3. E. B. Ford, *Genetic Polymorphism* (Cambridge, M.I.T. Press, 1965).

new mutations which are recessive to their normal alleles will stand a better chance of remaining a part of the genetic constitution (genome) of a population than if they were exposed to the full force of natural selection each time they arose.

When we combine this means of preserving variability in a population with the mechanisms of segregation, independent assortment, and recombination we can see that living species possess many mechanisms for preserving genetic variation. Add the influence that the activity of a gene at one locus may exert on one at another locus (a phenomenon called "pleiotropy" which is thought to be very common) and the net result is that, at the level of the genotype, contemporary sexually-reproducing species are the possessors of systems for generating nearly infinite variety. So it is that even though mutations are rare events, occurring in only one out of every 100,000 to one out of every million gametes, species are able to meet most environmental changes with at least some individuals possessing the genetic constitution to cope. But the point of distinguishing genotype from phenotype is that the variety available in a species is not limited to that already described but possesses yet another dimension, that of phenotypic variability. Here an example might well prove most instructive.

We are all aware that people seem to be getting taller. This is true in most parts of the world. In some areas, this so-called "secular increase in stature" has been going on for a century or more. In other areas it is a fairly recent event having taken place within the last generation. Some idea of the magnitude of this increase in body size can be gotten through inspection of garments, including suits of armor, used by Europeans of six or seven centuries ago. In most cases, their descendents are unable to get into the clothes of their ancestors with even the most determined effort. Some years ago, the reconstruction of La Scala, the famous opera house in Milan, ran into a snag when it was found that the contemporary Milanese could not fit their more ample pelves into seats which were presumably comfortable for their predecessors of 200 years ago. In the United States, average male stature increased by two inches between the First and Second World Wars and may still be increasing, although it appears to be leveling off. The most frequently purchased shoe sizes have steadily increased in both sexes, as has the demand for extra large shirts, underwear, outerwear and virtually all other garments except hats. Similar observations have been made in most of the industrial countries, perhaps most strikingly in Japan, where body size increases in both sexes have been noted just since World War II.

When such trends are recorded, particularly those limited to a single generation in time, the probability of genuine genetic change is quite low. Instead, it has been argued quite convincingly that the combined effects of improved nutrition, reduced incidence of infectious disease, and generally improved environmental conditions have led to the attainment of increased body size by contemporary human populations. Evidence to support this contention is available when comparison of the body sizes of individuals drawn from the more affluent segments of past populations are made with measurements of their less fortunate comtemporaries. Such comparisons, when based on conscription data collected in England,

France, and parts of Germany during the eighteenth and nineteenth centuries reveal differences of such magnitude that the more affluent young males seem almost drawn from some separate population, one characterized by significantly greater stature and weight. Since many of the larger, more affluent youths of nineteenth century England ended their university careers by going into the Anglican ministry, Stanley Garn has mused that "the benefits of clergy could be weighed in pounds and measured in inches."[4] If more evidence is needed on the role of nutrition and medical care in the determination of body size in humans, it is abundantly available in the records of statural increases and decreases during periods of intermittent famine in France and Germany during the eighteenth and nineteenth centuries.

Associated with these increases in body size has been an earlier attainment of sexual maturity in both sexes. The striking reduction in the age of occurrence of menarche in girls in both the U.S. and Europe during the last century (Tanner, 1968) has occurred during the same period that earlier attainment of adult body size has been observed. It remains to be proven that these phenomena are causally related, however. Nevertheless, these changes are of great significance from both biological and sociological points of view.[5]

If such important events are taking place in the absence of demonstrable genetic change, there must be nongenetic means of altering the phenotype. From the preceding examples it becomes clear that the people we measure, are the product of interaction between genetic endowment and environment with alterations in the environment producing significant differences in genetically similar individuals. The opposite is also true, i.e., we might expect, under the proper circumstances, individuals of different genetic makeup to resemble each other to a greater than anticipated extent. The evolutionary implications of the possible dissonance between the genotype and phenotype are very important. This is because natural selection operates on what the organism exposes to the environment, i.e., the phenotype, which may in many respects differ for a given genotype. The net effect of these additional extragenetic factors is to make the game of evolution more complicated, less predictable, and, to use Conrad Waddington's phrase, the "strategy of the genes" is infinitely more subtle than any existing genetic model is able to convey.[6] Nevertheless, such extragenetic events must be considered if any sort of understanding of the evolutionary process is to be attained. The maintenance of the variability required for adaptation to occur is itself an adaptation. As such it is the evolutionary product of natural selection. However, natural selection, operating as it does on the phenotype, will eliminate individuals, not genes, and some of the individuals who survive will do so because of their success in making suitable extragenetic adaptations. This is particularly true in species where behavioral traits allow effective adjustment to environmental stresses and thereby allow the accumulation of previously useless allelic variation at the loci concerned with traits

4. Stanley M. Garn and C. G. Rohman, "Interaction of Nutrition and Genetics in the Timing of Growth and Development," *The Pediatrics Clinics of North America* 13 (1966):353-379.

5. James Tanner, "Earlier Maturation in Man," *Scientific American* 218 (1968):21-22.

6. Conrad Waddington, *The Strategy of the Genes* (London: Allen and Unwin, 1957).

sensitive to such stresses. We will examine some examples of the interaction of our species with an environment altered by our own activities during the course of the succeeding chapters. But to illustrate the last-mentioned event we may turn to the example of survival of victims of diabetes mellitus in contemporary human populations.

This genetically transmitted trait is of serious consequence if not treated. The condition is caused by inadequate production of the hormone insulin in the pancreas. Lack of sufficient insulin in the blood results in a rise of blood sugar (glucose) levels. This happens because insulin induces cells to take up sugar, thereby holding blood glucose to physiologically tolerable levels. When blood glucose rises to critical levels, diabetic shock ensues, an event that is frequently fatal. Treatment of diabetic individuals through administration of insulin obtained from domestic animals has allowed most to live relatively normal lives. As a result, the diabetic now lives to reproduce in numbers equal to that of the unaffected segment of the population. Presumably, the gene or genes concerned with diabetic tendencies are now under reduced selective pressure and will, as a consequence, increase in number over a period of time. We may or may not wish to call the self-administration of insulin by the diabetic an adaptation, but the fact that his survival and reproduction will tend to produce an increase in the frequency of diabetes-mediating genes in the future may certainly be interpreted as evidence of adaptation of the genetic category, whatever the mechanism. If diabetes results from the presence of a single gene, changes in gene frequency are calculated directly through the medium of counting the affected individuals in the population. The ease of gene frequency analysis for a genetic trait is the greatest where the alleles at the affected locus can both be recognized, i.e., are codominant. When one allele is dominant in the presence of the other, the individuals (heterozygotes) possessing both will be indistinguishable from individuals possessing a double dose of the dominant allele (homozygotes).

If selection is operating to remove individuals exhibiting a trait coded by a dominant allele, it will affect every individual inheriting the trait and would presumably shift gene frequencies more rapidly than would be the case if the trait were coded by a recessive allele. In the latter case the only individuals exposed to natural selection are those for whom both alleles at the affected locus are of the recessive form. When we calculate the probability of occurrence of such homozygous recessive individuals in a population, we find that their number is equal to the allelic frequency squared. For example, in the well-known case of the genetic defect called phenylketonuria (PKU), the frequency of the deleterious recessive allele in the U.S. population is about 1 percent. As a result, the occurrence of affected individuals, i.e. recessive homozygotes in our population is: $\frac{1}{100} \times \frac{1}{100} = \frac{1}{10000}$. Despite the relatively common occurrence of the defective DNA segment in our population the loss of the PKU gene is an uncommon event. Such genes are shielded from selection sufficiently to remain in the population for a long time. How long and at what frequencies such genes will persist in a population is determined by the recurrent mutation rate producing new defective segments of DNA and the reduction in fitness experienced by their possessors. Methods for calculating levels of occurrence of such genes have been developed over the years. Examples of methods for handling all manner of cases can be found in other volumes of this series and in works such as Francis Johnston's *Microevolution of Human Populations.*

Under existing circumstances, it does not appear that diabetes or PKU are traits which enhance the adaptation of our species. It is not certain that in the case of diabetes at least, this has always been true. It is conceivable that what is now an unambiguous metabolic defect may have, under conditions of recurrent feast and famine, lent its possessors some small advantage in the maintenance of blood sugar levels. If such had been the case, the problem of diabetes could be viewed as the result of a shift in ecological relationships of the species.[7] Perhaps we will never know for certain if this is the case. It is difficult to hypothesize a selective advantage for the PKU defect at any time in the history of the species, although the future remains unpredictable. But, as will be discussed later, such traits may have a unique role in the shifting adaptation of the species.

The key question in any attempt to predict the future value of any genetic trait is "what will be the ecology of the future?" Since our environment will probably change in many unpredictable ways at a rate which is also unpredictable, our species, like any other, stands a better chance of survival if at least a few of even the most undesirable genetic traits are preserved in some sheltered state. We may then make the rather surprising observation that our species' adaptation involves the retention of a number of maladaptive traits. In consideration of a species' adaptation over long periods of time we must necessarily differentiate between the long-term "strategy" which can be discerned in the major events in the history of evolution and "tactical" measures by which a population manages survival in a given setting.

The analogy between "macroevolutionary" processes and a "strategic" overview and that of "microevolutionary" processes with a "tactical" application are useful ones for anyone attempting to understand the interaction of ecology and species' adaptability. Biologists have long been intrigued by examples of species which "won the battle but lost the war," often because of a too-successful genetic adaptation to a specific environmental setting. Usually such examples of extinction involve overspecialization, reduction in variability, and inability to cope with alterations in some aspect of the environment sufficiently well to compete effectively for an ecological niche. The narrower the ecological niche a species must rely upon for its survival the more precarious that species' adaptation. But effective competition for a given habitat requires that a species possess some advantage over other organisms which would displace it. This advantage is generally obtained through some form of specialization.

The tactics of evolutionary success involve the displacement of competitors in a habitat or the successful exploitation of a new habitat by making the necessary biological adjustments. Some of these ad-

7. J. V. Neel "Diabetes Mellitis: A Thrifty Genotype Rendered Detrimental by 'Progress'," *American Journal of Human Genetics* 14(1962): 353-362.

justments will be genetic, but during important transitional periods, extragenetic factors will perhaps be more important. From the species' standpoint an effective strategy is to have numerous populations employing different tactics in numerous areas, thereby allowing the species as a whole to "hedge its bets." There is some reason to believe that some of the most successful species have been those which hedged their bets most effectively. For many reasons, we must count our own species among those benefiting from a successful strategy. Some of the means by which a species may adapt to environmental stresses within the lifetime of the individual will be the subject of chapters three and four and the significance of these devices to the course of contemporary human evolution will be the topic of chapter five.

At this point in the general consideration of adaptation it is necessary to introduce some adaptive tactics that do not directly involve genetic change but are nevertheless significant factors in the success of the species and may translate into increased fitness for certain phenotypes. The genetic implications of such selection cannot be ignored just because they are harder to calculate. Following will be some arbitrarily chosen levels of adaptive response to convey some idea of the array of tactics our species might employ to survive in the face of environmental changes occurring over time spans too short to be dealt with by "descent with modification."

THE TACTICS OF INDIVIDUAL SURVIVAL: OR WHAT TO DO UNTIL THE PROPER GENOTYPE ARRIVES

Various definitions of adaptation have been offered over the years in an attempt to come to grips with the realities of the interaction of individuals with their environments. If we adhere to the most restrictive definition we will be confronted with the tautology that any living organism is "adapted" and therefore discussion of its relative adaptation becomes meaningless, adaptation being an either/or situation.

From the standpoint of the individual, the only true measure of adaptation is that of reproductive success. But this term is troublesome in that it only applies to relative fitness in the face of intraspecific competition for resources. What if intraspecific competition is in some way minimized for a time as through isolation of subpopulations subsequent to expansion into previously unexploited territory? The determinants of reproductive success may vary from population to population so that although differential reproductive success may be observed within populations, no one "best type" may be identified for the species as a whole. If the populations involved are mobile, the environment changing, or both, relative fitnesses of populations may change over time as well.

When populations occasionally share or exchange members, genetic adaptations occurring in high frequency in one population may be tested on an entirely different background in another. The result of periods of such exchange may, under the influence of natural selection, produce adaptive complexes which are widespread. Since they represent a medley of traits allowing effective adjustments to a number of environmental stresses, they may confer a capacity for adaptation rather than an adaptation in the strict sense of the word. This is true because the genetic potential underlying such traits is never fully expressed in any environment.

Therefore, individuals within such a species produce variable phenotypes while possessing similar genotypes when circumstances differ. They may also produce similar phenotypes while possessing different genotypes when environmental demands warrant it. From the species' standpoint, this is good strategy and will, in time, be genetically incorporated. We might view the process as adaptation through adaptability. This form of adaptation will probably involve many genetic loci and may employ the genetically encoded function of certain organs in different ways under different circumstances. How else can we explain the range of effects observable when certain secretions of the adrenal cortex increase in the circulation of their possessors? One of these ketosteroids can be shown to reduce inflammation in tissues, alter the cellular storage of blood sugar, and enhance the breakdown of protein molecules to produce more blood sugar (gluconeogenesis). These same hormones may at certain stages in the life of the individual, delay growth, and, if present in high concentrations long enough, cause reduced adult size. In addition, the presence of these anti-inflammatory glucocorticoids appears to suppress the immune response and alter the function of the liver in several important ways. In other words, any stimulus producing a vigorous adrenocortical response will have many and widespread effects on the organism as a whole. Some of these effects are potentially damaging but are tolerable for short periods when an emergency must be dealt with, which is usually the case.

This so-called stress response is seen to be similar in a number of mammalian species. Much of what we know about its characteristics has been learned through animal experiments. But confirmation through measurement of human physiological changes has been possible in recent years. The course of the response to stress, called the General Adaptation Syndrome (G.A.S.) by Hans Selye,[8] has been studied in a number of species, yielding important information concerning how we cope with stress. Some of this information has been used to develop effective clinical procedures to assist humans whose stress responses are either inadequate or inappropriate. The G.A.S. is only one of a number of physiological changes which have been identified when individuals undergo stress. Those changes, occurring within the lifetime of the individual, may change his phenotype, sometimes temporarily, sometimes permanently, but do not alter his genotype.

Of immediate interest for the present discussion is the degree to which phenotypic alteration may occur. Since there are a number of levels at which phenotypic alteration may be manifested, the whole system may be regarded as a graded response system in which the biological investment of the individual in dealing with a stress situation is held to a level commensurate with the level and duration of the stress encountered. The operational meaning of this concept will be clearer after the examination of some characteristics of certain levels of the graded response system employed by humans under stress.

Developmental Acclimatization

When an organism is exposed to en-

8. Hans Selye, *The Physiology and Pathology of Exposure to Stress* (Montreal: Acta Inc. Medical Publishers, 1950).

vironmental stressors throughout all or some significant portion of its growth and development, it may take on physiological or morphological characteristics which remain with it for life and become a permanent component of the adult phenotype. The depth of such alterations may be great. The earlier the changes occur, provided the stimulus for change is sufficiently strong and prolonged, the more fundamental the phenotypic alteration. Perhaps the most impressive display of the potential for such change is seen in experiments with a number of species (not including man) in which appropriate environmental alterations can produce phenotypes which are sexually opposite their genotypes. Sex reversals can be produced by suitable hormone administration when the embryo is in the earliest stages of sexual differentiation and the degree of reversal obtained is commensurate with the earliness of its exposure to the relevant hormone.[9] It is thought that human intersexuality has, in some cases, resulted from accidental or pathological alteration of hormone levels during critical stages in morphogenesis, as when excessive amounts of androgenic steroids are secreted by the mother suffering from tumors of the adrenal cortex or the anterior hypophysis.

In general, the earlier an individual is affected by abnormal circumstances, the greater will be the long-term effects. This is particularly true during the critical early stages of cell proliferation (hyperplasia) and differentiation marking the embryonic period. In humans, embryonic growth occupies the first three months (trimester) of the gestation period. As a rule, serious trauma encountered during embryonic life are sufficiently disruptive to cause termination of the pregnancy. This might be considered an adaptive response on the part of the mother. A sufficiently defective embryo will develop into a nonviable fetus. If a nonviable fetus is carried to term and then lost, the effect is to withdraw a fertile female from the reproducing segment of the population for an extended period.

The placental system of humans effectively cushions the embryo and fetus from most shocks (see chapter 2) but some shocks, for instance, certain disease organisms, hormone imbalances, toxic materials (including carbon monoxide), and some nutritional deficiencies can affect the *in utero* environment sufficiently to alter the path of development. In some cases, reduced size of a full-term newborn will result. Sometimes premature birth with its attendant hazards occurs. If the underlying problem is not corrected, some stresses may persist through the perinatal, neonatal, and infant stage of life, in which case a number of phenotypic alterations are possible. When these changes are effected by early exposure to stress and become an irreversible condition reflected in the adult phenotype, we may legitimately consider them "developmental acclimatizations." They are useful and *irreversible*.

Acclimatization[10]

When an individual is exposed to a stressful circumstance, it is possible to ad-

9. John Money, "Effects of Prenatal Androgenation and Deandrogenation on Behavior in Human Beings," in *Frontiers in Neuroendocrinology*, eds. William F. Ganony and Luciano Martin (New York: Oxford University Press, 1973).

10. The distinctions made in this section follow those recommended in: G. Edgar Folk Jr., *Introduction to Environmental Physiology* (Philadelphia: Lea and Febiger, 1966).

just in a less pervasive way than that occurring in developmental acclimatization, but still exhibit measurable phenotypic changes of varying duration. Some such changes or acclimatizations may persist for an extended period as when a permanent move to a significantly different climate occurs, or may be of shorter duration as experienced when seasonal changes occur. The significant factor in such changes is that they are useful (adaptive) and *reversible.* Even such phenotypically obvious changes as muscular hypertrophy belong in this category. In a real sense, they provide examples of an unexpressed genotypic capacity being exploited in the face of a new set of demands and might be viewed as a temporary reordering of priorities, which may persist as long as the conditions evoking it persist.

Acclimation

It is also possible to respond to a single short-term stress with physiological alterations of relatively modest dimensions. Such responses will utilize resources already present to correct an imbalance which lowers functional efficiency and may be perceived as discomfort of some form. For instance, most sea level dwellers experience a variety of distressing symptoms if suddenly transported to a high altitude environment. The range of complaints expressed by various individuals stressed in this way includes headache, fatigue, shortness of breath, dizziness, nausea, hypersensitivity to alcohol, and sometimes a bloody nose and the inability to sleep. In most cases, these symptoms subside after a few days as a result of physiological adjustments which occur in a generally predictable sequence. In most cases, people quite naturally reduce their activity level and involuntarily increase their respiratory rate. As a result, changes occur in blood chemistry. These changes involve the amount of carbon dioxide present in bicarbonate ions and a resultant alteration in the acidity of the blood. Blood *pH* shifts to normal levels after a time, partly because extra red blood cells appear in the circulatory system thereby increasing the oxygen transport capacity of the blood. This reduces the need for hyperventilation and restores the normal physiological state of the individual. Many systems are involved in this form of adaptation and if the degree of oxygen deprivation is too severe and prolonged, the system of an unacclimatized individual may be overwhelmed. The results can be fatal. In many cases, the capacity to endure such stress is increased by repeated exposures, in which case a state resembling acclimatization may be attained.

Habituation

Shorter term responses to stressors might be grouped under the heading of habituation. This might be viewed as a reduction in the level of physiological response to stressful stimuli. As such, it permits the organism to maintain a normal homeostatic state despite potentially disruptive stimuli and, in a sense, cushions the individual from adverse secondary reactions to his own response mechanisms. For instance, if the hand is plunged into cold water, the physiological response is one of vasoconstriction, one which reduces the flow of blood into the extremity and thereby protects against the lowering of the body's core temperature. If persistent, such vasoconstriction may, and sometimes does, permit the temperature in the extremities to fall to a level where tissue damage (frostbite) occurs. Habituation desensitizes the sensory apparatus sufficiently to allow circulation to be maintained in the extremities at a lower temperature than before. "Specific habituation" refers to reductions in pain experienced in a specific region, as in the chilling of a single toe, while "general habituation" would refer to reduction in the overall intensity of a physiological response, such as vasoconstriction in the entire periphery when the chilling of a single toe is experienced. The important distinction between habituation and acclimation is that habituation is concerned with the prevention of damage to the organism resulting from its own overreaction to a stressful circumstance, while acclimation refers to a graded adjustment to the stress itself.

Behavioral Adaptation

There are a number of ways to deal with stress at the behavioral level thereby reducing the need for even moderate and temporary physiological adjustments. Some of these behavioral means lie at the threshold between physiological and strictly behavioral phenomena, as in the case of reflexes. If you touch the top of a hot stove, you will be treated to an example of a human reflex action as you involuntarily draw your hand back from the pain-inducing stimulus. There are, of course, better ways of making such observations, many of which have been highly ingenious. Such reflexes are seen in a wide variety of organisms with perhaps some of the most complete studies having been done in invertebrates. Such reflexes are clearly adaptive but relatively inflexible. Some of the undesirability of this sort of programmed response can be perceived when we contemplate the consequences of stepping on a hot rock while walking barefoot on a narrow cliffside path. Fortunately, this stereotypic behavior is relatively uncommon in adult humans and is usually evoked by genuinely noxious stimuli and is therefore of continuing adaptive value.

Other nonstereotypic behavior may also be directed to avoidance of potentially damaging stimuli. All sorts of learned responses may be employed by humans to reduce the hazards of their environment. Simple acts, like putting on a coat when cold, fall into the category of adaptive behavior, as do many complex acts which we perform with relatively little mental effort. Humans, as individuals, draw upon a rich repertoire of learned behaviors, many of which may be employed to adapt to changing conditions. The greatest value of such behavioral responses is that they may be used in a wide variety of changing circumstances and altered at the discretion of the user. New responses may be produced to deal with situations encountered for the first time and plans may be made to deal with circumstances which may arise in the future. A full treatment of the variety and significance of behavioral responses in human interaction with the environment is impossible in the present context but it is certain that the responses constitute a large and important component of the strategy of adaptedness characteristic of our species. When the behavior of individuals is combined to deal with the environment and when those individuals must learn to plan and coordinate their activities, the net effect of group effort is enhanced out of all proportion to the sum of the efforts of the individuals involved.

So in consideration of human adaptability some mention must be made of cooperative behavior, even though a comprehensive presentation of the case for inclusion of group cooperation in the list of traits subsumed in the category of man's biological adaptations deserves a separate, lengthy treatment.

Group Cooperation

Humans cooperate to propagate. More than that, they cooperate in some way in almost every aspect of their lives. This is true around the world and is a trait which has probably characterized our species for several million years. As is discussed in chapter two, our primate ancestry has predisposed us to adapt through collective activities. When a species is made up of individuals who exhibit great variability in both genotypes and phenotypes, the combination of traits possessed by any one individual is small compared to that possessed by the species as a whole or, even to that possessed by the breeding unit, or population. However, if the population has learned to cooperate, the individual abilities of all of its members may be coalesced to confront the environment with a full array of talents which in their aggregate, constitute the new adaptive phenotype of the corporate entity. As long as the benefits gained from cooperative activities outweigh the disadvantages of group life . . .contagion, interpersonal conflicts, and the like, populations adapting in this manner will experience an advantage over those which do not. All other things being equal, populations which possess effective cooperative behavior will expand in number and range faster than those lacking it. It does not matter whether the capacity for such behavior is genetically transmitted or not once a system for transferring the necessary learning from generation to generation is developed. Whatever genetic traits enhance the effectiveness of such an adaptation will be the beneficiaries of a superior fitness because of their association with it. Since the net result of this form of adaptation is increased flexibility in selecting means to deal with the demands of a changing environment, it might be viewed as adaptation through adaptability. As cultures have become increasingly complex along with their increased capacity to buffer their possessors from the stresses of the natural environment, many aspects of the evolutionary process have been altered. This is a topic which will concern us in the final chapter of this book.

The mere fact that we are here on this planet to discuss the course of our own evolution in the face of uncounted millions of species extinctions since life began, is testimony to the success of our species' strategy. But our species has undoubtedly paid the price of this strategy through the loss of populations which found themselves unable to cope with the demands of their environment. Depending upon where we draw the species line in our evolutionary history, we may infer some such inability in certain of our ancestors. We know that we are in many respects different than our ancestors of a million years ago. We probably differ from much more recent ancestors as well. The world has changed and we have changed with it. Since these changes are the product of the species' strategy being played out in tactical responses to environmental demands, the nature of our ecological niche is of considerable interest. So this is perhaps a proper point to assess the position among living things that we as a species occupy. To use the hallowed term chosen by Huxley in his nineteenth-century effort to describe our biological affinities, let us proceed to a consideration of "man's place in nature,"[11] while adding the twentieth-century caution that it is woman's place too.

For Further Reading

Eaton, Theodore H. Jr. *Evolution.* New York: W. W. Norton, 1970. A readable and comprehensive treatment of the biology of adaptation. Suitable for the student seeking a general background.

Grant, Verne *The Origin of Adaptations.* New York: Columbia University Press, 1963. A monumental compilation of current evolutionary theory. A bit difficult for a beginning student, but full of information.

Ross, Herbert H. *Understanding Evolution.* Englewood Cliffs, N.J.: Prentice-Hall, 1966. Useful for assimilating the "big-picture," the background, and mechanisms of evolution written for the student with minimal biological training.

Stebbins, G. Ledyard *Processes of Organic Evolution.* 2d ed. Englewood Cliffs, N.J.: Prentice-Hall, 1971. Well-written and accurate, this work provides a host of useful examples for considering the mechanisms of evolution.

Wallace, Bruce and Srb, Adrian *Adaptation.* 2d ed. Englewood Cliffs, N.J.: Prentice-Hall, 1964. Probably the best overall point of entry for the beginning student of evolutionary theory.

Williams, George C. *Adaptation and Natural Selection.* Princeton, N.J.: Princeton University Press, 1966. Definitely not for the uninitiated, but recommended as the most complete coverage of arguments concerning the place of altruism and group selection in the adaptation of populations.

Bibliography

Lederberg, J. 1966. "Experimental Genetics and Human Evolution." *American Naturalist* 100:519-531.

McKusick, Victor A. 1969. *Human Genetics.* 2nd ed. Englewood Cliffs, New Jersey: Prentice-Hall, Inc.

Muller, H. J. 1967. "What Genetic Course Will Man Steer?" In *Proceedings of the Third International Congress of Human Genetics (September).* Edited by James F. Crow and James V. Neel. Baltimore: Johns Hopkins Press.

Sinclair, David. 1969. *Human Growth after Birth.* London: Oxford University Press.

Stern, Curt. 1973. *Human Genetics.* 3rd ed. San Francisco: W. H. Freeman and Co.

11. Thomas H. Huxley, *Evidence as to Man's Place in Nature* (London: Williams and Norgate, 1863).

2 The Species *Homo sapiens* In Taxonomic Perspective

We might well consider our adaptation as being composed of a number of traits and potentialities superimposed upon a background characterized by the very general necessities for the perpetuation of life itself. Since life as we know it is a product of events occurring on this planet, our most basic needs reflect the peculiarities of our terrestrial home. For instance, certain elements universally essential for all known life are present here on earth in greater abundance than in the universe as a whole. Carbon, oxygen, and nitrogen are among the elements found in relative abundance in our environment but are relatively scarce in the total cosmos. Table 1 illustrates the quantities of the life-associated elements on earth with their general availability in the universe. From these quantities it is clear that life is not the sort of phenomenon which would be expected to emerge spontaneously at numerous places in the universe, but instead bears the characteristic imprint of the planet earth. At the same time, the planet certainly bears the imprint of life. The presence of life on this planet has altered its characteristics in fundamental ways. It is probable that the very presence of life in the form of green plants has changed the balance of gases in the earth's atmosphere in a manner which would make the events which originally gave rise to life's appearance impossible. So it is probable that the origin of life as we know it was, at least on this planet, a one-time and one-time-only event. It is possible, of course, that life in some form arose elsewhere in the universe but it would, in all likelihood, be quite dif-

TABLE 1

ELEMENTS IMPORTANT TO LIFE THEIR OCCURRENCE ON EARTH AND IN THE UNIVERSE AT LARGE (PERCENTAGES OF TOTAL MASS)

Element	*Plants*	*Earth's Crust*	*Universe*
Oxygen	59	49	0.06
Carbon	21	0.1	0.03
Hydrogen	16	3	87
Nitrogen	3	0.0001*	0.008
Potassium	0.1	0.1	0.000007
Calcium	0.1	2	0.0001
Silicon	0.1	14	0.003
Magnesium	0.04	8	0.0003
Phosphorous	0.03	0.07	0.00003
Sulfur	0.02	0.7	0.002
Sodium	0.01	0.7	0.0001
Iron	0.005	18	0.0007

*Atmospheric Nitrogen is a highly significant element, making up 78 percent of earth's atmosphere.

ferent from the forms we are accustomed to.

We share basic requirements with all living things. The requirements of cellular metabolism are quite similar among plants, animals, and even bacteria. The energy transfer processes basic to life use similar compounds in the cells of plants and animals. For instance, certain polyphosphate compounds, including adenosine triphosphate (ATP) serve as devices to lock up energy for future consumption in the cell. Cells synthesize and store similar compounds, including fats, carbohydrates, and proteins using similar biochemical pathways. In most cases, the methods by which these compounds are degraded to release energy are similar. In most cells, metabolic reactions are catalyzed by a peculiar class of proteins called enzymes. If we scan the various forms of life with particular attention to the structure and function of enzymes we will find a high degree of structural similarity among those enzymes performing similar functions in very dissimilar animals.[1] When cells of protozoons, pear trees, and people make proteins, they do so by a complex coding process whereby the sequence of amino acids formed is regulated by the DNA found in the nucleus of the cell. There are certain ubiquitous, small compounds involved in metabolic processes such as, some of the B-complex vitamins, the flavinoids, the carotines, heme groups, isoprenoid compounds, and iron sulfide. Often, these compounds act in conjunction with proteins as coenzymes or as the catalytic centers of enzymes themselves.

The large number of characteristics we share with other living things means that we share many of their requirements and limitations as well. This existence of shared requirements and limitations also implies that certain evolutionary experiments are just not feasible. In many respects, the list of potential experiments grows shorter as evolutionary events produce species with increasingly more complex requirements. For instance, since the metabolic activity of green plants has released large quantities of molecular oxygen in the earth's atmosphere, many organisms, ourselves included, have become dependent upon oxidation reactions in cellular metabolism. The presence of abundant oxygen (nearly 21 percent of the atmospheric gases) has become so important in our species that an interruption of less than ten minutes in respiratory activity is generally fatal.

Taxonomy, the study of the relationships of life forms, is importantly concerned with the discovery of the basic requirements shared by various species as well as with their morphological similarities. Some highly significant findings concerning these relationships have emerged from contemporary studies of molecules important to the lives of a number of species. In many instances, the relationships discerned through examination and comparison of protein molecules have been in close agreement with relationships ascertained through anatomical resemblances, the study of the life cycles, and ecological relationships of species. This is, of course, as it should be. The ideal for development of a classification scheme reflecting the interrelatedness of life forms has long been to produce a natural classifi-

1. For a number of comparisons see: M. D. Dayhoff, and R. V. Eck, *Atlas of Protein Sequence and Structure* (Silver Spring, Md.: National Biomedical Research Foundation, 1969).

TABLE 2

MAN'S PLACE IN NATURE
AS VIEWED THROUGH BIOLOGICAL CLASSIFICATION

Kingdom	Animalia
Phylum	Chordata (Animals possessing notochords)
Subphylum	Vertebrata* (Animals possessing backbones)
Class	Mammalia*
Infraclass	Eutheria*
Order	Primates*
Suborder	Anthropoidea (Monkeys, apes and humans)
Superfamily	Hominoidea* (Apes and humans)
Family	Hominidae (Hominids and prehominids)
Genus	Homo (Living and extinct humans)
Species	sapiens (Contemporary man and direct antecedents)

*Levels of the taxonomic hierarchy singled out for detailed discussion with respect to human adaptation.

cation rather than one developed around certain traits (called taxonomic keys). Such keys, although easily identifiable, often have little evolutionary significance. Modern classification systems are in a very real sense reconstructions of evolutionary history. They are based on as much information as can be obtained concerning the environmental factors which have caused species to diverge and pursue separate evolutionary courses. It is accurate to say that a contemporary classification scheme is an evolutionary statement and, as such, tells us much about the adaptations the species has exploited to survive and compete for its place in the sun. Table 2 is an abbreviated classification scheme presented to provide one perspective on our place in the scheme of things. We will have occasion to refer to certain levels of this scheme as important elements of our species' adaptation are discussed.

As can be seen in Table 2 we are members of the primate order, are mammals, and are (terrestrial) vertebrates. Although each level in the classification table has its own significance, we have chosen to concentrate on the significance of being a vertebrate, a mammal, and a primate in the understanding of the interaction with the environment characteristic of our species. Within this context, it will be possible to consider most of the limitations and opportunities which have shaped the adaptation which we recognize as uniquely human.

THE IMPORTANCE OF BEING A VERTEBRATE

Being vertebrates, we possess considerable mobility. We are capable of controlling our movements and to go where our needs can be best satisfied. Perhaps most importantly, we can inhabit the land areas of the world. Living on land is more challenging than residence in an aquatic habitat, but it offers more opportunities as well. The need to counteract gravity is much more demanding on land than in the water. Testimony to this fact is the existence in the sea of the largest of living animals. The engineering problems organisms as large as whales would encounter in a dry land environment are formidable. Since the mass of an organism is increasing faster than its surface area as the animal gets larger, great size ultimately produces an intolerable load on the weight-bearing limbs. As a rough example of the principle involved, we might visualize what would happen if humans were to double in stature. If such increase in size was proportional the body volume of the average person would increase to approximately the third power. The weight increase associated with the volumetric expansion would require considerably more massive limbs if locomotion were to be possible. A comparison of the limb structure of an elephant with that of a pig will give some idea of the structural alterations which might result. The massiveness attendant to such size increase would, in short, necessitate a chain of adaptations if survival is to be achieved. The limitations on size are much greater on land than in water and we, as terrestrial vertebrates, are shaped in part by the physical laws governing the relationships of mass to movement.

Many other physical laws come into play when the determination of body size is effected. Fluid dynamics, the speed of conduction of nerve impulses, and compressional and tensile strength of biological building materials are all components of the equation defining the limits of body size in a specific environment. When we add to these factors, the energy demands encountered as size increases, a factor to be considered separately in a later section, it should come as no surprise that successful terrestrial vertebrates exhibit a relatively narrow range of body sizes when compared to marine forms. Certainly no animals are free of restrictions on size regardless of their environment, but our ecological niche, involving as it does occupation of a dry land habitat, has imposed a clear set of limitations on the size individuals of our species will attain. It should be pointed out that we are near the upper end of the size range for terrestrial vertebrates and, as a result, are capable of supporting a relatively large brain.

By definition, vertebrates are animals with backbones. Most vertebrates have many other bones as well. The combination of bones and skeletal muscles has evolved into an elegantly effective set of levers which employ the force generated by the shortening of muscles to do all kinds of work. The gross movements we perform when we walk, run, or carry objects are the product of this evolutionary process. But so is the intricate interaction of muscle, nerve, and brain characteristic of the artist or craftsman. The exquisitely subtle movements involved in many human activities are the product of the mutual evolution of bone and muscle as a system. Facial expressions are also the product of this evolution and so is the form of communication we call speech. In short, much of what we consider characteristically human about ourselves is a part of our vertebrate heritage. But inevitably, all advantages are acquired at a cost. Some of the costs we incur have already been mentioned. They are, in large part, associated with greater metabolic cost for survival than would characterize invertebrates. The demand on the environment for nutrients, oxygen, and minerals characteristic of terrestrial vertebrates is generally transmitted into a restless search for the necessities of survival. In a very real sense the possibility of increased activity creates a need for it. As living forms become more complex they gain the freedom to prowl about the world, expanding their ranges and, by encountering many new environmental conditions, experience new selective pressures. It is not surprising then that the evolutionary process has been very active among vertebrates, particularly terrestrial vertebrates. But it is also true that as vertebrates gained more freedom to explore new habitats, they did so at the cost of increased need to supply their metabolic demands. So it is that vertebrate evolution has set trends which are significant in shaping the course of human adaptations. This is particularly true of the most active classes of the vertebrate phylum, the birds and the mammals. Consideration of some elements of our mammalian heritage will help us further define our place in nature.

THE IMPORTANCE OF BEING A MAMMAL

Despite the fact that the largest living mammals are found in the sea, mammals are a product of the evolution of terrestrial vertebrates. Marine mammals invaded the sea at a relatively recent date and are still required to get oxygen from the atmosphere much as we do and, as a consequence, seals, whales, and dolphins are all capable of drowning. The secondary invasion of the sea by mammals may be viewed as an indication of success of the mammalian adaptive complex which allowed certain forms to penetrate habitats previously inaccessible to them. This phenomenon, "adaptive radiation," appears to occur when evolution produces an adaptive innovation enabling its possessors to increase in number and expand their range relatively rapidly. Viewed from the perspective of evolutionary strategy, adaptive radiation allows the testing of new ecological niches by populations which apply whatever tactics they possess to cope with the demands encountered. When micro-differentiation between populations occurs under such circumstances, it may produce new species. When many new niches are penetrated, many new species may arise. This appears to have happened when mammals displaced the reptiles as the dominant form of life on earth toward the end of the Mesozoic Era, about 100 million years ago. By this time vertebrates had been evolving for about 400 million years and had occupied terrestrial areas of the planet for at least 300 million years, first only as amphibians and later as increasingly well-adapted reptiles culminating in the great radiation of the dinosaurs. Many of the characteristics which define the mammalian class were to be found in some of the later forms of dinosaur.

A. S. Romer,[2] who has devoted a long and distinguished career to the study of vertebrate evolution, has offered the opinion that certain therapsid dinosaurs had achieved most if not all of the adaptive characteristics we attribute to present day mammalian forms and therefore might have occupied a taxonomic position between the Class Reptilia and the Class Mammalia. Whether or not the therapsids occupied this position, some form of reptile/mammal did and there is little doubt that mammals are the product of natural selection working on certain reptilian species to produce new adaptive complexes. More important than the origin of mammals for the present discussion is the trait complex that has made the mammals so successful, keeping in mind

2. Alfred S. Romer, *The Procession of Life* (Cleveland: World Publishing Co., 1968), pp. 227-243.

the fact that mammalian affinities are of central importance in opening up the opportunities and defining the limitations of human adaptation.

The word "mammal" is derived from the possession of mammary glands, modified sebacious glands which secrete milk. Mammals nurse their young, a trait of great importance in both physiological and behavioral adaptation, as will be discussed below. Another characteristic of mammals is the possession of hair or fur. According to Morris,[3] the possession of hair and mammary glands is the result of a closely linked set of adaptations. To paraphrase Morris' argument, during the evolution of hair, which was a part of a temperature-regulating complex, certain skin glands appeared. Hair provided the animal with warmth while the secretions of the glands kept it cool. The mammary glands developed from some of these glands. Therefore, an insulating fur coat probably preceded the acquisition of mammary glands. Viewed in this manner, the origin of mammals would be more directly the result of the attainment of a temperature-control system than of any other selective process. So we might view mammals as being most importantly defined as animals which control their internal environment within well-defined limits.

Birds have solved the problem of temperature regulation in a similar manner, using feathers rather than hair or fur as insulation. Since birds do not have hair, however, they do not have the glands associated with it and have not, as a consequence evolved mammary glands. Some interesting variations within the Class Aves have produced certain traits which come close to the mammalian pattern as may be seen in the production of "crop-milk" by pigeons and the existence of "preen glands" in a number of birds.

The combination of sweat glands and fur allows the mammal control of body temperature in the face of wide fluctuations in external temperatures. This is particularly valuable on land where the surrounding medium, air, varies greatly in temperature over the course of a year or even a day. As a consequence of temperature regulation (homeothermy) mammals are able to remain active during both day and night hours as well as throughout the year. Also, the range of habitats open to organisms with temperature-control systems is considerably greater than for those without. Mountain, temperate, and arctic environments can be exploited once the problem of keeping the internal temperature constant is solved. Not only is the range of mammals increased over that of forms without temperature regulation, but also it is possible for them to rely upon more sensitive physiological systems. Therefore, mammalian metabolism is highly efficient and mammals are, as a result, more active, energetic animals.

With increased activity and more forms of physiological regulation comes the necessity for a reliable circulatory system and, associated with it, an efficient respiratory system. The demands inherent in the coordination of high activity levels with a finely tuned set of physiological controls call for the possession of a highly developed central nervous system with special emphasis on improvements in the brain. The mammalian heart avoids the mixture of oxygen-rich blood leaving the lungs, on the way to the brain and systemic

3. Desmond Morris, *The Mammals* (New York: Harper & Row, 1965), pp. 9-20.

circulation, with the oxygen-depleted venous blood returning from the tissues. This means the oxygen supply available to sensitive tissues can be held quite constant and that metabolic reactions can proceed at a constant rate despite fluctuations in activity levels or external conditions. The brain is the greatest beneficiary of this system for supplying high-quality, oxygenated blood and, as a consequence, has been able to expand and differentiate greatly in a number of mammalian species.

A part of the complex involved in the supply of oxygen-rich blood to the tissues is the muscular diaphragm which mammals possess. This network of muscular tissue is located between the thoracic and abdominal cavities. Its presence allows mammals to obtain more oxygen quickly than would be possible without it. The diaphragm, through its capacity to enlarge the space surrounding the lungs, employs the creation of a negative pressure gradient to induce the flow of outside air through the airways to the lungs. In the lung oxygen diffuses across moist membranes to capillaries where it is attracted to red blood cells. Red blood cells carry oxygen to tissues where it is needed, release it, and pick up carbon dioxide produced by cells as a byproduct of metabolic reactions. Carbon dioxide is released in the lung and the red blood cell is ready to begin the cycle again. The process is a closely coordinated one in which the circulatory, respiratory, and cardiac elements all work in close harmony. Neat, efficient—but expensive in terms of the demands such a system makes on the nutrient supply.

The demand for nutritional intake to supply this expensive, sophisticated system makes it necessary for the organism to be able to eat, breathe, and maintain activity levels all at the same time. In order to do this, it is essential that the ingested food be prevented from entering the lung where it would interfere with the exchange mechanism for oxygen and carbon dioxide and probably damage the thin, moist tissues involved. In mammals, this separation is effected by the presence of a secondary palate, located between the breathing passage and the swallowing one. So we, as good mammals, have a trachea through which we inspire and expire air and an esophagus, through which food and liquids travel on their way to the stomach. Unfortunately, the separation of the two systems is not entirely fool-proof, as is painfully obvious when we swallow something while inhaling and particles enter the airway. This is not only uncomfortable but potentially dangerous, since the air supply can be interrupted causing unconsciousness and possibly death. Even when foreign matter passes through the trachea, its presence in the lung can cause serious problems including pneumonia. The occurrence of such unpleasant events should not be too surprising when we consider the close proximity of the structures involved and the necessity of their synchronous activity. Perhaps more remarkable is the relative rarity of accidents of this sort, a tribute to the effectiveness of the multiple adaptive processes which have characterized the evolution of mammals.

As mammals became increasingly active with improvements in their circulatory, respiratory, and nervous systems, selective pressure developed favoring improvements in the limbs. Four-footed (tetrapod) reptilian ancestors had gotten along with limbs which, in many cases, sprawled out to the sides of the body. This is all right if rapid running is not a necessary activity, but speed of locomotion is more efficiently obtained when the weight-bearing structures, the limbs, are positioned beneath the mass they support. Evidence of experimentation in this direction is seen in the skeletons of some of the dinosaurs, but a truly effective solution is a trait developed by mammals. This involves reorientation of the limbs toward the body's midline by bending the forelimb so that the "elbow" protruded rearward and the hindlimb so that the knee protruded forward. We humans have undergone some additional rotations in our hindlimbs as a part of our specialization for walking upright (erect bipedalism). You can get a good idea of the benefits of the mammalian limb positions if you compare the strain you feel when doing push-ups compared to that felt when you swing your elbows under you and cheat a little. Of course, repositioning of the limbs, makes a number of other structural alterations necessary, particularly in the limb girdles, which now must absorb the shock of increased activity. Also, the force generated in rapid locomotion must be greater for longer periods and to account for this, certain muscles must enlarge while others become less important and may therefore reduce in size and in some cases divide and perform finer, more precise actions.

The alternations in relationships and the attainment of greater precision in certain movements is necessarily reflected in both the peripheral and the central nervous system. Balance becomes more important as are the sensory monitors of position and the portions of the brain integrating motion and position. Some of the changes involved in the central nervous system facilitating rapid locomotion have led to alterations in the shape and size of the brain itself. Mammals have taken a different evolutionary direction than birds in this respect. As anyone who has watched a bird flying rapidly through a thick grove of trees can testify, birds have evolved an amazingly efficient system of coordinating muscular activity and sensory input. In the avian system however, there is a predominance of the balance-controlling part of the brain, the cerebellum. In mammals expansion of the cerebral areas predominates and in certain species including ourselves, the cerebral cortex has undergone the greatest expansion. Many of the differences between the brains of organisms are associated with the locomotor patterns they pursue while others reflect predominance of certain special senses in the organism's habitual activities. In any event, the improvements in mammalian locomotion have been reflected in improved central nervous systems and, as more neurological pathways are cross-linked the very character of the brain is altered. Where quantitative change ends and qualitative change begins is not easily determined, but we, as a species, have clearly been the beneficiaries of both. So changes in locomotor patterns must, for our purposes, be an important mammalian innovation. More will be said on this point in the discussion of primate characteristics.

Another aspect in which mammals improved their adaptation over that inherited from reptilian ancestors is in the development of specialized teeth along the tooth row. These teeth enable the organism to make the nutrients contained in ingested food available in a very short time. Again, high activity levels, an ever-demanding brain, and the control of body temperature combine to make the mammalian system one which must be fueled at regular intervals. While it is true that most mammals can live off their fat for a time, an event seen most frequently in hibernating and estivating forms, dietary deprivation is not only uncomfortable, but if sufficiently prolonged is dangerous. Devices which make feeding more efficient and more economical will certainly improve the fitness of organisms experiencing heavy energy demand. In mammals one of the identifiable adaptations improving the fueling operation is the possession of a "heterodont" dentition, one in which teeth are shaped differently and perform different functions in different parts of the tooth row. There are a number of variations on the heterodont dentition theme among mammals, with carnivores showing one form of adaptation, and grazing ungulates another. In carnivores, the meat-eater's diet is best handled by grasping canine teeth and slicing, chopping premolars and molars of a specialized form called carnassials. In the grazing animals where vegetable material must be ground down and wear is a problem, the molars are very high and have complicated occlusal surfaces. We have a combination of shearing incisors, puncturing and grasping canines, and premolars and molars with relatively complicated grinding surfaces. As our dentition would indicate we can eat a wide variety of foods, i.e., are omnivorous, and can therefore exploit a wide variety of nutritional resources. This dietary versatility is reflected in the extensive range our species has been able to occupy and may be considered also as a part of our mammalian heritage.

Another mammalian characteristic which bears great significance to the way in which organisms interact with their environments is the production of live young. Mammals are not the only organisms which reproduce by internal fertilization and give birth to young which have spent a considerable length of time growing inside the mother. This also happens in some sharks, snakes, and bony fishes. But it is most widespread among mammals and the time *in utero,* the gestation period, attains by far its greatest length among mammalian species. Not all mammals give birth to live young, the monotremes (duck-billed platypus and the spiny anteater) lay eggs in a manner very similar to that seen in reptiles. The monotreme and reptile "amniote" egg possessing an amnion, chorion, allantois, and yolk sac, also seen in mammalian development, gives good evidence of what kind of system preceded the most prevalent mammalian one. Mammals possess a complex deciduous organ called the placenta. Marsupials are an infra class of mammals which nourishes the newborn in a pouch on the ventral body wall after a very short gestation period. The process in marsupials involves the attachment of the very young infant to a nipple inside the pouch where it remains while achieving a state of development which will enable it to survive independently. This is clearly a more intimate association of mother and offspring than that seen in monotremes, but is much less intimate than the true placental (Eutherian) association found in most mammals, ourselves included.

In placental mammals, the placenta establishes an intimate connection with the uterine circulation. In our species, there is an actual implantation of the structure in the uterine wall. As the embryo develops, the placenta grows rapidly and provides a membrane across which substances may pass in both directions. The placental membrane is very selective. It will not allow many molecules to pass at all, and others are only permitted to pass when broken down to their constituents, as is the case with fats and a number of proteins. Other substances, oxygen, carbon dioxide, and water among them, cross the membrane freely following a concentration gradient. Some substances cross the membrane only by the expenditure of energy (active transport) but do get where they are needed even when it is necessary to move them against a concentration gradient.

Among mammals, there are several varieties of placentae categorized according to the degree of intimacy achieved between the fetal and maternal circulation. Since the fetus has many individual characteristics which are different from those of the mother (remember, only one-half of its genes are acquired from either parent), it is a foreign body to the mother's immune system. If there were a free admixture of fetal and maternal blood, the maternal immune system would eventually react vigorously to attack and destroy it. The only way in which this disastrous event could be prevented would be to either deactivate the maternal immune response, a prohibitively risky alternative, or to shield the maternal immune system from the stimulation which would follow detection of the foreign antigens of the fetus. The different forms of placentation observed in Eutherian mammals represent a series of compromises in the attainment of efficiency, as would result from free admixture of fetal and maternal blood on the one hand, and protection from unwanted and destructive immune response to the presence of fetal antigens on the other. The compromise struck in our species involves the pooling of maternal blood alongside the chorionic membrane of the fetus. It is not a foolproof system. Some fetal-maternal incompatibilities are known to result in aborted fetuses and the loss of newborn infants who have been attacked by maternal antibodies (as in *Erythroblastosis fetalis*, the so-called "Rhesus" incompatibility). If anything, human placentation is a compromise tending to favor efficiency while running the risk of occasional losses to fetal-maternal incompatibility. Some rodents have an even more intimate association of fetal and maternal blood, but most mammals tend toward less intimate contact.

The retention of the developing embryo and fetus in the uterus raises the probability of its survival after birth. There is ample time to develop the complex interactions between the several systems involved in the regulation of the internal environment. Since mammalian homeostasis is more complex than that of species which do not regulate body temperature, the neurological apparatus is also more complex. More monitoring devices are essential. Rises and drops in body temperature must be detected promptly as must alterations in oxygen and carbon dioxide levels which could produce destructive changes in metabolically active tissues, including an expanding brain. Coordination of all of these systems calls for a large and complex brain which, in turn, adds to the demands on the life-support system. An extended period *in utero* allows these elaborations to take place while the developing organism is insulated from environmental shocks and predation. In a very real sense, we see in mammals an emphasis on quality as opposed to quantity as the focus of the reproductive process. Fish may lay millions of eggs to produce one viable breeding pair. Amphibians and reptiles are a bit more economical as are birds, but in mammals, there is a real trend toward the production of offspring which have a relatively high probability of survival.

The mammalian trait of possessing mammary glands, combined with the tendency to produce less young at any one time, leads to another characteristic of mammals differentiating them from most other organisms. This is the pattern of parental care of young animals after birth. At first this continued association of parent and young is a purely nutritive function but it generally extends well beyond such rudimentary elements into the area of parental protection, training, and guidance. The periods of parental care and guidance vary greatly among mammals as does the degree of neurological maturation at the time of birth. But it is, in general, much longer than the time allotted to such activities in nonmammalian forms.

Associated with an extended period of postnatal dependence and learning is a delay in the onset of sexual activity. Again, there is considerable variation among mammals with respect to the time of sexual maturity but it is among mammals that the interval between birth and reproductive activity is the greatest. This is, of course, consonant with the aforementioned trend toward reproductive parsimony and may be translated into terms of longer generation time in mammals. Appropriately, the delay in onset of reproductive activity is associated with a general increase in longevity in mammals as compared to most, but not all, nonmammalian species. Tortoises and parrots are rather obvious exceptions to this rule.

The extended period of immaturity, coupled with parental care and guidance, allows the young mammal to experiment with elements of the environment in a manner inconceivable to an organism preoccupied with a daily round of activities focused on the single question of survival. As a result, most young mammals indulge in some form of play. On the face of it, this sort of frivolous expenditure of time, energy, and parental patience would seem to be maladaptive, a misallocation of resources which will always be in short supply. How, then, can play behavior be explained, particularly in view of its prevalence among mammalian species? Certainly from a tactical standpoint two competing species, one which produced young who got right down to business and the other cursed with playful offspring would appear to be unequally matched in the game of survival. Here again, however, tactical considerations are tempered by strategic ones. Play is an effective way of learning, of developing new tactics without running the risks of the "real world."[4] Models of real situations are developed and suitable patterns of behavior are developed to cope with them. Thus, learning is facilitated, and situations which are crucial but infrequent may be simulated with the result that successful behavior is developed under the watchful eye of the parents and losses of young are minimized when survival-threatening situations arise, as always they must. Of course, this system is most easily perpetuated where litters are small and the young can be the beneficiaries of individual parental attention. In this way, tendencies toward repro-

4. Bernard Campbell, *Human Evolution* (Chicago: Aldine, 1966), pp. 48-54.

ductive parsimony are reinforced. Production of fewer young, with quality assuming greater importance than quantity in the reproductive process, has, over time, produced a characteristic and highly successful strategy among the Class Mammalia. Integral elements of this strategy are learning and some degree of cooperation between individuals of the species. With the lapse of a greater length of time between birth and reproduction, generation time has increased. An important concomitant of this increase in generation time is that genetic experiments will be more costly, since the replacement of any loss due to suboptimal genotypes will take longer and the population runs a greater risk of extinction. As a result of selection against populations whose tactics have involved primarily genetic adaptations to environmental changes, mammalian species which have had the capacity to adapt through extragenetic mechanisms have been at an advantage. Homeostatic devices characterize mammals. So, the foundation for adaptation by extragenetic mechanisms was laid early in the history of the class. With the added dimensions of improved neurological integration, learning, modeling, and cooperation, the mammals have been set firmly on the path toward evolution of organisms characterized by high quality and endowed with an evolutionary strategy that might well be characterized as adaptability. The obvious importance of this mammalian characteristic will be the subject of much of the succeeding discussion.

Since the adaptation of our species is so much a product of the opportunities and demands inherent in being a mammal, it is worthwhile at this point to list the traits which have been the concern of this discussion, several of which will be referred to in the accounts of specific human adaptations. As should always be the case, these traits, while frequently cited as taxonomic keys to aid in the identification of a specimen and its taxonomic affinities, are more important as indicators of the relationship of the organisms involved with their environments. In other words, while couched in anatomical, physiological, and behavioral terms, this trait list is really a comprehensive statement of ecological relationships and the adaptations to them characterizing the Class Mammalia. Table 3 is such a trait list.

TABLE 3

TRAITS CHARACTERISTIC OF MAMMALS

1. Presence of mammary glands
2. Possession of hair or fur
3. Internal control of body temperature
4. Possession of heterodont dentitions
5. Possession of a single dentary bone (mandible)
6. Live birth of young (except in monotremes)
7. Tendency toward reduced litter size
8. Extended period of postnatal parental care
9. Elaboration of central nervous system
10. Generally high level of mobility and activity
11. Relatively late onset of sexual activity (long generation time)
12. Play behavior
13. Sociability and in certain cases, cooperation
14. Emphasis on learning rather than stereotypic behavior
15. Relatively long life span

THE IMPORTANCE OF BEING A PRIMATE

Primates are, in a number of respects, "primitive" mammals. They evolved from the basic insectivores which are believed to have been the earliest mammalian forms and which are also believed to have been predominantly arboreal (tree-dwelling) animals. Perhaps more important than any so-called primitiveness of insectivore ancestors is the probability of their arboreal habits. Certainly, successful adaptation to life in trees involves peculiar demands on the special senses, central nervous system, and the organs of locomotion. In addition, insect eaters were clearly able to digest animal protein. But vegetable foods are abundant in arboreal habitats. This makes it possible to adapt to a combination of both animal and vegetable nutrient resources and gain considerable flexibility in the dietary repertoire. Since the strategy of adaptation through adaptability is an important element of primate success, such versatility stands a very good chance of having had an important influence on the course of one stage of human evolution.

The problems of locomotion in an arboreal habitat place a special burden on the function of the visual sense while rendering the olfactory sense less effective. Leaping from one tree branch to another is, under the most favorable circumstances, a hazardous undertaking. To do so without sharp visual acuity, good depth perception, and color discrimination is simply too hazardous to be undertaken. Much of the importance of primate affinities to the formation of human adaptive strategy will derive from the successful exploitation of the arboreal habitat. In an effort to reconstruct the sequence of events in the formation of the assemblage of traits so important to our success as a species, it will be useful to turn to some living primate species which may serve as models of stages of our own evolution. In so doing, we find that even the most distantly related primate species, those of the Suborder Prosimii, yield important clues to the origins of some of our most characteristic adaptations.

Most prosimian species give the impression of being quite different from what might be called "typical primates," in an anatomical or behavioral sense. One form, the tree shrew (Tupaia) has been included among the Prosimii by some primate taxonomists and excluded from the Primate Order entirely by others.[5] A separate taxonomic category, intermediate between the Primate and Insectivore Orders has been proposed by others. Whatever the final disposition of the taxonomic status of the tree shrew, its adaptation has interest for students of human adaptive strategy. Tree shews are arboreal, nocturnal, and insectivorous. Immunological data give support to the argument favoring their inclusion among the primates, but in many respects their anatomical characteristics bear little resemblance to what would be expected in a primate. They have long snouts, their eyes are directed laterally, and, in most respects, they have more resemblances to rodents and, of course, insectivores, than to primates. Living in trees, eating insects, and depending upon a well-developed sense of smell, these organisms might well represent a current model of the transitional stage between insectivores and primates. If we turn to some of the other contemporary prosimians, we see evidence of a trend toward increasing reliance on vision with a reduction in the length of the muzzle and a movement of the eyes to a point where overlapping visual fields are present, as in the tarsiers. Also among the prosimians, locomotion by leaping from branch to branch has be-

5. Joseph Birdsell, *Human Evolution* (Chicago: Rand-McNally, 1972), pp. 191-195.

come a highly developed trait. This involves a shift of the center of gravity to a point somewhere in the pelvic region, in most cases involving the presence of a well-developed tail which functions as a balance organ.

With the shift of the center of gravity to a point over the hindlimbs, leaping can be accomplished without tumbling and the animal can depend upon landing on the hindlimbs. Grasping, five-digited (pentadactyl) appendages are a relatively primitive trait that is put to good use by tree-traveling tarsiers and leaping lemurs (not to mention the lovable, loafing loris). Branches are grasped and held, often by the hindlimb while in a sitting posture made comfortable by the presence of the center of gravity in the pelvic area. Sitting upright with the grasping forelimbs freed for experimentation and feeding behavior, some living prosimians allow us to see how the freeing of our ancestor's forelimbs might have occurred. Once the opportunity for improved manipulation using the forelimb presented itself, a number of advantages could be secured through its exploitation. An important development was use of hands to secure food which was inaccessible before, as in stripping the cuticle from certain forms of fruit. This increased the variety of foods available. Precise movements of the hands arising from improved hand-eye coordination became an important adaptation. Hand-eye coordination in manipulation of objects requires improvement in the neurological control over the organs involved and integration of visual and tactile input with motor control. The complexity of this combination of demands is something the mammalian brain is capable of accommodating. With repeated exposure to the demand for such activity and the increased fitness an appropriate response would confer, the evolutionary process favored increasing reliance upon an improved central nervous system. Since the improvements most favored in these circumstances are characterized by flexibility in their application, it may be said that the prosimian grade of organization already exhibits some of the most crucial elements of the adaptive strategy humans have adopted. So we might consider the freeing of hands, through refinement of leaping and sitting behavior in an arboreal habitat an aspect of our heritage possessing great antiquity.

Large families are not desirable for tree-dwelling species. The hazards of rearing young in an arboreal habitat make the occurrence of multiple births a trait subject to strong selective pressure. Reductions in litter size, mentioned earlier as an important characteristic of the mammals, takes on additional importance in tree dwellers like prosimians. Contemporary prosimians generally produce their young one at a time. The benefits of small litter size, individual attention and increased learning opportunities for young mammals, are therefore conferred on young prosimians generously. In addition, the females of several prosimian species experience only a single ovulation each year, so the young are spaced at wide intervals. This spacing enhances the effect of a single birth by assuring a fairly lengthy interval during which the young prosimian is the beneficiary of his mother's undivided attention. Learning through imitation and instruction is thereby facilitated and will include the employment of improved hand-eye coordination with emphasis on the associative capacity of the enlarging brain.

Most of the contemporary prosimians are not characterized by elaborate social organization. In many species, there is a preference for solitary nocturnal activity and evidence of territorial marking. We do not know if such traits characterized any of our ancestors, but it is clear that, if our ancestors were antisocial, at some point in the evolutionary process, the benefits of social interaction began to manifest themselves. In any event, all of the contemporary anthropoids exhibit some form of social behavior. In addition, they have retained and in many ways, elaborated upon the traits which arose in a prosimian-like ancestor, improved brains, wider choice of nutritional intake, and great flexibility in the use of the hands.

Most contemporary anthropoids still are arboreal, but there are some important exceptions, particularly among the Old World monkeys. Several species of baboon have developed a primarily terrestrial mode of existence, as have several forms of macaque. In some cases, sleeping in trees at night and foraging in open country during the day makes up the daily round of activity, as is seen among baboons in East Africa. In India, some macaques have taken to village life, often to the consternation of the taxpayers. Some macaques have established residence at surprisingly high altitudes in the Himalayas. Reports of populations living above the 4000 m level give evidence of considerable expansion of range beyond that occupied by a tropical, tree-dwelling ancestor. Living apes also exhibit a range of habitats betraying a capacity to exist outside of forested areas. In all cases, the imprint of an arboreal adaptation remains. We may see the evidence of this adaptation in our own biological characteristics. Somewhat paradoxically, it is likely that an arboreal adaptation was an important prerequisite for the attainment of the erect bipedal form of locomotion so important to our contemporary adaptive complex and is, in most of its elements, so uniquely human.

There is a reasonable basis for the contention that our species descended from ancestors who used their forelimbs for locomotion at least part of the time in their forest environment. Swinging from branch to branch (brachiation) can be an effective means of getting from place to place in a densely forested area. Contemporary gibbons are extremely skilled at this sort of locomotion although neither they nor any other primate may be properly classified as full time brachiators. It would not have been necessary for our ancestors to have been full time brachiators for them to have gained a significant advantage from the possession of anatomical traits making such locomotion feasible. Grasping digits, capable of alternately holding and releasing branches as hand-over-hand progress is made through the trees, would have been a part of the anatomical apparatus inherited from their ancestors. Possession of a loose wrist joint which could support the weight of the organism while rotating would be essential and could have developed quite simply from the loose fitting wrist of a prosimian-like form. Once such a wrist and forearm did develop it became possible to rotate the hand independently of upper arm motion, in the movements of "supination" and "pronation" (palm-up vs. palm-down position of the hands effected by the rotation of one of the long bones of the forearm, the radius around the other, the ulna). We use our capacity to supinate and pronate in many ways, most of the time taking it so much for granted that we are scarcely conscious of its performance.

Another anatomical feature important to an arboreal primate experimenting with brachiation is a loose shoulder joint; one which would allow the hand to be held directly over the head to suspend the organism and permitting sufficient freedom of movement to allow full rotation of the arm at the shoulder joint ("circumduction" in proper anatomical terminology). Such a shoulder joint would rely heavily upon the action of surrounding muscles to keep the joint together, particularly under the strain involved when swinging rapidly from branch to branch while supported by only one arm at any given moment. We have such a shoulder, strong, freely rotating, and heavily dependent upon muscles and ligaments to maintain structural integrity. This dependence upon muscles to hold the human shoulder together can be appreciated when the effects of certain forms of muscle and nerve damage or paralysis are seen. In such cases, the proximal long bone of the arm, the humerus, may fall over an inch away from its normal point of articulation with the glenoid fossa of the scapula. The muscles of the shoulder are massive, strong, and well positioned for the work of raising the arm, supporting the body, and pulling it along while suspended. The powerful flexor muscles of the human arm may still be trained to perform this function very effectively. One of them, biceps brachii is a favorite recipient of special treatment inducing hypertrophy followed by display in males of the contemporary human species.

In addition to these anatomical characteristics, brachiation requires increased stability of the joint between the sternum (the breastbone), and the clavicle (collarbone). This joint forms the ventral anchor of the pectoral girdle and is importantly involved in the transfer of force from the forelimb to the trunk and, when brachiating, the transfer of the weight of the rest of the body to the forelimb. Involved in this anatomical adaptation is the development of a robust clavicle with sufficient room for a number of muscle attachments. Increased importance in weight transfer is reflected in greater breadth and general massiveness of the sternum. Comparison of a human sternum or that of one of the great apes with the sternum of a nonbrachiating Old World monkey will illustrate how far changes in the sternum have gone in the direction of increased structural rigidity.

One of the characteristics associated with increased sternal breadth is a general broadening of the thorax in relation to its anterior-posterior depth. Humans, along with brachiating apes, have broad, shallow chests. Evolution of this type of chest necessitates changes in a number of skeletal features as well as alterations in the positions of muscle origins and insertions and a general repositioning of a number of internal organs. Since suspension from tree limbs causes the trunk to be carried vertically a good deal of the time, the force exerted by gravity is also felt in a different way. This requires changes in the way internal organs are held in place by connective tissues. Also, the spinal column is used in an entirely different way when the organism spends much of its time in an upright position. In essentially four-footed, leaping monkeys, the lower end of the spinal column, the lumbar region (that area of the spine between the rib cage and the pelvic girdle), is long and flexible, making the use of the spine as a springlike propulsive organ possible. In brachiators, a stiffer spine is better adapted to the upright position and the lumbar region is shorter. Of course, the weight transfer function of the spine operates in an entirely different plane in the upright position. Instead of the weight of the body being transferred through the limbs by way of an arched spinal column in a manner analogous to that seen in the structure of a suspension bridge, weight is transferred up and down the spinal column itself, through the bodies of the vertebrae. This is a factor of considerable importance when the positioning of the center of gravity must also be controlled as in the attainment of erect bipedal, rather than a suspensory mode of locomotion. It is necessary, with the possession of a broad, flattened thorax and rearrangement of the viscera of the trunk, to develop a number of curves in the spinal column. These curves place the center of gravity over the pelvis. In this way walking becomes possible. This in turn is associated with repositioning of the head and its articulation with the spinal column.

In a quadriped, the head gets wherever the organism is going first. With the spine approximating a position parallel to the ground, and with the head at the forward end of the spine, the special sensory apparatus (eyes, nose, ears, mouth) face forward. This means that the eyes and the point of entry of the spinal cord into the skull are approximately opposite each other. The weight of the skull creates a tendency for it to drop off the end of the spinal column. This tendency is countered by the presence of muscles inserting on the rear of the skull and originating on the spinous processes of the vertebrae. If the sense of smell is important and feeding activity requires a large dentition, the muzzle, the part of the skull extending farthest forward, will be large and heavy. As a result, reliance on the sense of smell and large dentitions require more heavily developed neck muscles. Often these nuchal muscles insert near or at the very top (vertex) of the skull and in some cases are so robust that they require the presence of raised crests of bone to accommodate their attachment.

The combination of reductions in muzzle size and the attainment of semi-erect posture allows a reorientation of the skull. In our species, it is balanced on the top of the spinal column with no massive muscle attachments on the outside of the cranial vault. The point of entry of the spinal cord is at the base of the skull (the foramen magnum) at about a 90^{0} angle from the line of sight of the eyes looking straight ahead. The precise role this repositioning has played in allowing or inducing expansion of the brain is a matter of speculation. But the form of the braincase has certainly been influenced by the balancing of the skull on top of the spinal column. At the same time, additional demands are made on the spinal column, since compressional forces are now exerted on the vertebral bodies as a result of the weight of the skull being transmitted through them. A full discussion of all of the anatomical ramifications of the attainment of erect posture is beyond the scope of the present discussion, as is that concerned with expansion of the brain to its present dimensions.[6] However, it should be clear from the foregoing that much of what constitutes our anatomical adaptation can be traced to our affinities to arboreal primates. We walk erect rather than swing through the branches of trees,

6. For an excellent discussion of the anatomical changes involved see: Campbell, *Human Evolution*, pp. 86-106.

but in many ways, the two modes of locomotion are similar. The anatomical changes involved in the shifting of the spinal column into a vertical position serve both well. Since we have good reason to believe that arboreal habitats came earlier in our evolutionary history than erect bipedalism, it is logical to assume that adaptations which proved advantageous in forest-dwelling primates were subsequently exploited in a nonforest environment. This gave rise to a primate possessing a very potent combination of traits now including an enlarged brain, with enhanced freedom to expand, an improved neurological system, and hands free to use in a wide range of functions, locomotion being relegated to the lower limb. Learning, play behavior, and gregariousness, all characteristic of anthropoids would, when combined with these new advantages, open a wealth of opportunities. Acquisition of tool use and manufacture and improved communication through vocal signals culminating in language further enhanced their value. Other advantages possessed by our ancestors at this stage in human evolution would have been associated with a general condition of adaptability, a lack of overspecialization to any single ecological niche and a capacity to exploit many sources of subsistence.

It will be recalled that prosimian primates give evidence of a close relationship to insectivores. Some prosimians are insectivorous while others appear to be primarily vegetarian. Others tend to be omnivorous. A similar range of dietary preferences can be seen throughout the primate order, with most primates favoring a vegetarian diet but indulging in some animal protein when the opportunity arises. There is a strong likelihood that human evolution went through a phase during which hunting and meat eating were important behavioral traits.[7] It is thought that cooperation and communication among groups hunting large animals made these traits important for the survival of their possessors and that there was strong selective pressure favoring individuals and populations endowed with the capacity to develop these traits most effectively.

The capacity to learn, to socialize, and to communicate is part of a general primate heritage. Exploitation of these capacities in a new environment opened new opportunities. These in turn, created new pressures which would be translated into increasing refinement of what was to become an evolutionary strategy of adaptation by adaptedness. The capacity to develop and maintain some form of culture became an integral part of this adaptation and has had the effect of drastically redefining the ecological relationships of the species. No other species has achieved the capacity to alter its environment to the extent humans have. As will be discussed at some length in chapter five, the effects of culture as a buffering system between ourselves and the natural environment may be shaping the course of contemporary human evolution in ways we are not yet able to predict.

Our heritage of adaptations as vertebrates, mammals, and primates has defined the tactics with which human populations adapt to environmental stresses which have influenced the course of past evolution. Much of our response to stress is automatic and a large proportion of our adaptation is genetically determined. But

7. William A. Stini, "Evolutionary Implications of Changing Nutritional Patterns in Human Populations," *American Anthropologist* 73 (1971): 1019-1030.

the strategy employed by our species has, because of the unique background of human evolution, additional tactics to bring into play, often preventing other tactical responses but usually working in concert with them. For this reason, while we may learn much through observation of and experimentation with other species, the only suitable subjects for the study of the relationship of ecology and human adaptability are humans. Although study of human subjects presents many difficulties, its value for the understanding of our current place in the natural world and our potential effect on our habitat make it essential to continue to ask the questions which have occupied students of human biology since its emergence as a discipline. Such studies are not always popular and the goals of research on human evolution and variation are frequently misunderstood. But much remains to be learned about our relationship to the natural world and it is possible that by learning more about the tactics employed by human populations exposed to specific forms of environmental stress, we may be forewarned of the effects of human activities on the future environment of the species.

It took a long time to produce the adaptive complex characterizing our species. Evolution is possible because environmental changes generally occur gradually. We are now capable of effecting drastic alterations in our habitat in a short period of time. As a result, we run the risk of exceeding the limits of human adaptability. Much of the study of contemporary human biology is concerned with the definition of those limits by viewing human interaction with the environment in an evolutionary perspective. Next, we will turn to some examples of the effects of changing environments on human populations.

For Further Reading

Birdsell, Joseph B. *Human Evolution*. Chicago: Rand McNally, 1972. A general introduction to physical anthropology with a distinctly ecological approach. One of the best places for the beginning physical anthropologist to sort out his interests, although he will have to go elsewhere for real depth of treatment.

Calvin, Melvin *Chemical Evolution*. New York: Oxford University Press, 1969. A useful collection of information on the events which made life possible. The ultimate in informitarian arguments emphasizing the place of predictable chemical reactions in the evolutionary process.

Campbell, Bernard G. *Human Evolution*. 2d ed. Chicago: Aldine Atherton, 1974. The structural-functional approach to human biology is treated better here than in any other single work. Readable for a student with modest background in biology.

Griffin, Donald R. and Novick, Alvin *Animal Structure and Function*. 2d ed. New York: Holt, Rinehart and Winston, 1970. An excellent source for the student seeking understanding of how organisms function and what the differences between taxonomic groups tell about their adaptations.

Keeton, William T. *Biological Science* 2d ed. New York: Norton, 1972. A useful general reference for any student in the biological sciences. Although a textbook in general biology, its organization and indexing make it a handy, quick reference for assimilation of concepts.

Morris, Desmond *The Mammals: A guide to the Living Species*. New York: Harper & Row, 1965. A useful reference for the student interested in the taxonomic relationships and natural history of mammalian species.

It makes available a comprehensive bibliography of modern taxonomic theory.

Romer, Alfred S. *The Procession of Life*. Cleveland: World Publishing Co., 1968. A statement on the evolutionary relationships of all life by an eminent vertebrate palaeontologist writing for the general reader.

Sagan, Carl *Life, Article for Encyclopedia Britannica*. New York: Encyclopedia Britannica Inc., 1970. A general treatise on the origins and limitations of life as a product of conditions on the early earth. Heavy going for a beginning student but crammed with useful information.

Young, J. Z. *An Introduction to the Study of Man*. Oxford: Clarendon Press, 1971. A wide-ranging study of human biology by an eminent vertebrate biologist. Most worthwhile to the student with some biological background to draw on.

Bibliography

Hulse, Frederick S. 1971. *The Human Species*. 2nd ed. New York: Random House.

Jolly, Allison. 1972. *The Evolution of Primate Behavior*. New York: The Macmillan Co.

Kelso, A. J. 1970. *Physical Anthropology*. Philadelphia: J. B. Lippincott Company.

Lasker, Gabriel W. 1973. *Physical Anthropology*. New York: Holt, Rinehart and Winston, Inc.

Pilbeam, David. 1972. *The Ascent of Man*. New York: The Macmillan Co.

Romer, Alfred S. 1969. *The Vertebrate Body*. 4th ed. San Francisco: W. B. Saunders Co.

Simons, Elwyn L. 1972. *Primate Evolution*. New York: The Macmillan Co.

Simpson, George G. 1953. *Major Features of Evolution*. New York: Columbia University Press.

————. 1961. *Principles of Animal Taxonomy*. New York: Columbia University Press.

Weichert, Charles K. 1965. *Anatomy of the Chordates*. 3rd ed. New York: McGraw-Hill Book Co.

Young, J. Z. 1962. *The Life of Vertebrates*. 2nd ed. New York: Oxford University Press.

3 Genetic Adaptation
Evidence of Darwinian Evolution in Human Populations

THE ROLE OF ADAPTATION IN THE ORIGIN OF SPECIES

Charles Darwin had a sophisticated and often misunderstood view of the mechanisms of evolutionary change. He took the long view, being willing to accept a time frame of many millions of years for evolution to produce known forms of life. The process was a uniformitarian one in which unremarkable events combined in a series to produce remarkable results. His philosophical commitment to this view of life may be traced to his interest in uniformitarian theories of the origin of land forms which had been developed by the geologists Hutton and Lyell. However, Darwin was handicapped by lack of knowledge in two important areas, earth history and genetics. His difficulties in the area of earth history derived in part from the intolerably short time scale (24 million years) for the history of the planet Earth favored by the followers of Lord Kelvin. Kelvin's calculations of the age of the earth were supported by convincing mathematical deductions based on the rate of cooling of the planet's crust after separation from the sun. The calculations themselves would not have been misleading if they had not been based on faulty assumptions.[1] As it turned out, the age of the earth is more than 100 times greater than the estimates favored during Darwin's day. Darwin was not able to reconcile the pace of evolution he postulated and the variety of species which had evolved with the age of the earth as calculated by Kelvin. This difficulty caused him to modify some of his views later in his career thereby weakening his theory in certain respects. If he had incorporated the elements of Mendelian genetics in his theoretical framework, it is possible that some of the defects which crept into his latest work could have been avoided. But it seems certain that Darwin remained unaware of the crucially important concepts of hereditary transmission published by Mendel in 1865. It was not until after the beginning of the twentieth century that Mendel's work was rediscovered, leading to the birth of the science of genetics, and another quarter century was to pass after that before the science of genetics and evolutionary theory were to be combined in what has been called the modern "synthetic" theory of evolution.

1. For a full discussion of Darwin's difficulties see: Loren Eiseley's *Darwin's Century* (New York: Doubleday, 1961).

This is not to say that Darwin has been rejected in any sense, however. In view of the difficulties he faced in the formation and defense of his theory, we can only marvel at the accuracy and sophistication of his insights concerning the mechanisms of "descent with modification." His view of the origin of adaptations was essentially "preadaptive," i.e., whatever favorable variation appeared was the result of some process which caused accidental alterations of characteristics. If these accidents proved beneficial, they would be preserved in the species, giving their possessors an advantage in the competition for scarce resources and mates. The result was that such favored individuals were represented in subsequent generations in greater proportions than their less-favored conspecifics. Quoting Darwin himself from *On the Origin of Species by Means of Natural Selection* (1859): "The preservation of favorable variations and the rejection of injurious variations, I call Natural Selection . . .every slight modification, which in the course of ages chanced to arise, and which in any way favoured the individuals of any of the species, by better adapting them to their altered conditions, would tend to be preserved. . . ." There can be no doubt that in 1859, Darwin saw evolution as the operation of selection on chance variation. If he had had knowledge of some mechanism for producing that variability requisite for selection to operate, it would have been possible to fill in some of the gaps in the theoretical framework which caused him so much difficulty. We have the advantage of knowing how mutations occur and the means by which natural selection can change the genetic constitution of a population over time. However, well-documented examples of the entire process of adaptation are still not plentiful.

One of the better known and widely cited examples of adaptation in the presence of recognizable selective factors involves the changes in phenotypic frequencies among peppered moths of the industrial midlands of England. In this now-familiar example, dark forms of the moth increased in frequency as their coloration protected them from predation by birds during a period when the trees upon which the moths roosted were blackened by air-borne soot. Since predating birds found the "wild-type" peppered moths more visable on soot-blackened trees, natural selection favored the dark moths. Records kept over 100 years by local lepidopterists confirmed the greater frequency of the dark moths in recent years. There are reports that current efforts to clean up the air in England's midlands have met with considerable success and that more and more trees in the area now present the pattern of light, lichen-dotted bark which prevailed during the preindustrial period. It might be expected that a reversal of selective pressure would occur from this change and there is evidence that a decrease is now occurring in the frequency of the dark phenotype. We have in this series of observations support for the preadaptive view of evolutionary change within species. Experimental data to support this view has also been made available through the work of the Lederbergs with bacteria, using a technique called "replica plating."

The Lederberg experiments involved growing cultures of Escherichia coli, a common intestinal bacteria, on agar. A number of agar plates were cultured from the same parental strain of E. coli, i.e., cloned, so that as far as possible, they would possess the same genetic potential. Next, some colonies were introduced to agar plates impregnated with the antibiotic streptomycin. Most of the bacteria exposed to streptomycin quickly died. Some, however, continued to grow and ultimately produced cultures which were entirely streptomycin-resistant. By going back to the original plate yielding the resistant colony, it was shown that subsequent colonies drawn from it also possessed the quality of streptomycin-resistance, as did the descendants of the resistant strains with a few exceptions.[2] These experimental results, revealing as they do that the trait needed to survive in the presence of an antibiotic was part of the genetic potential of the species before it had experienced the effects of the antibiotic, may be interpreted as indication that such traits occur by some process other than as a response to a specific stress. In other words, evolution may be regarded as being preadaptive in the sense that traits which will prove to be adaptive arise before there is any such advantage in their possession.

Since no trait may be regarded as neutral in the eyes of natural selection, preadaptive traits might therefore be, to some extent, maladaptive up to the time they become adaptive. As a corollary of this we must conclude that the species maintains its requisite variability at a cost. The cost may vary from trait to trait and from species to species and may be difficult to estimate in many instances, but logic compels us to stipulate its presence. Since all living species continue to pay the price of maintaining variability in this manner, it must be considered an integral part of the general strategy of evolution. With this in mind, we may argue that "genetic load," the burden of deleterious genetic traits borne by all species may be viewed as "evolutionary capital." This is not to say that we would benefit from conditions which would increase our genetic load, all of the evidence is to the contrary, but it is likely that the history of life has been such that the production of new variability, the event we call mutation, is an integral part of the life process. Let us now turn to an example of genetic adaptation in humans, an example that is of value for its completeness to all evolutionary biologists.

HUMAN GENETIC ADAPTATION: ALTERATIONS IN THE HEMOGLOBIN MOLECULE

Our mammalian ancestry has presented us with many opportunities as well as raising the energy cost of survival (for discussion, see chapter two). Many organs and organ systems are involved in the satisfaction of our requirement of a high metabolic rate. The cardiovascular and respiratory systems function as a unit to provide essential oxygen to the organs and tissues where oxidative metabolism is the most important source of energy. The actual transport of oxygen is the function of the red blood cells. If we were forced to rely on the amount of oxygen soluble in the blood to provide our metabolic requirements, the limits imposed would make life as a mammal impossible.

So we can see that the function of the human organism is greatly dependent upon the functions of the red blood cell. The primary function of the red blood cell is the synthesis and transport of hemoglobin which has the important property of attracting oxygen in an oxygen-rich environment and releasing it in an oxygen-

2. A detailed account of these and similar experiments may be found in R. P. Levine, *Genetics*, 2d ed. (New York: Holt, Rinehart and Winston, 1968).

poor one. Also, hemoglobin attracts carbon dioxide where it is in high concentrations and releases it where it is in low concentrations. This means that the red blood cells passing through the pulmonary capillaries where oxygen is in high concentrations and carbon dioxide in low concentration will pick up oxygen and release carbon dioxide, a state during which its hemoglobin is mostly oxyhemoglobin. In the capillaries of the peripheral circulation, where the carbon dioxide produced as the result of the oxidation of carbohydrates is in high concentration and oxygen in low concentration, hemoglobin releases oxygen and picks up carbon dioxide, becoming carboxyhemoglobin. The mechanism underlying the capacity of hemoglobin to perform these functions involves the changes in oxidation state of an iron atom found in the center of a porphyrin ring which is a part of the hemoglobin molecule.

Our hemoglobin, like that of all mammals and most vertebrates is a four chain molecule.[3] The chains orient themselves in a characteristic manner giving the molecule a shape permitting it to move its iron-bearing portions (the heme groups) closer to or farther from each other. This enhances the oxygen-carrying and releasing capacity of the molecule. Its location in the wall of the mature red blood cell allows the hemoglobin molecule to be distorted by forces distorting the cell wall itself as during passage through the capillaries. Red blood cells have the shape of biconcave discs. Their shape might be visualized as resembling that of a thin doughnut with a membrane stretched over the central hole. In passing through the capillaries, red blood cells are oriented so that their diameter fills the interior of the capillary. The diameter of the red blood cell is such that it just fits through a dilated capillary. Being oriented the way it is, a red blood cell passing through a capillary is subjected to greater pressure in its thin central area than in its thicker peripheral region. The peculiarities of fluid dynamics in such circumstances (laminar flow in a cylindrical space), arise from friction which retards flow near the walls of the capillary. The rate of flow is greater in the center of the vessel and the red blood cell, being thinnest at its center, the very point where fluid pressure is greatest, is distorted, its central portion being pushed forward, creating a bulge. Distortion of the membrane results in alteration of the relationships between the chains of hemoglobin within it, facilitating release and pickup of oxygen and carbon dioxide. However, this system also leads to stresses which are potentially destructive to the red blood cell.

Perhaps this repeated exposure to stress is responsible for the relatively short life of the human red blood cell which, on the average, lives only 120 days before it is destroyed and its components recycled. In certain circumstances, red blood cells live considerably less than their alloted six score days. One condition which drastically reduces the life span of the red blood cell is the condition called sickle-cell anemia which has been shown to result from an alteration in the protein sequence in a part of the hemoglobin molecule.

When the hemoglobin is altered in a certain way, there is a tendency for the cell to distort in a maladaptive manner. When viewed through a microscope many such cells appear to have taken on a crescent or

3. For a full exposition of the characteristics of hemoglobin see: V. M. Ingram, *The Biosynthesis of Macromolecules* (New York: Benjamin, 1965).

sicklelike form. This usually happens when the oxygen saturation of the blood is low as occurs during heavy exercise and under conditions of hypoxia as at high altitude. Sickled cells are not capable of performing their oxygen transport function and are usually destroyed.

Most destruction of red blood cells is done in the spleen. When the spleen becomes congested by an overload of sickled red blood cells, it becomes nonfunctional as well, aggravating an already dangerous situation. When this situation has gone far enough to damage the spleen and restrict the tissues' oxygen supply below critical values, the results are fatal. Death generally occurs early in the lives of sufferers of sickle-cell anemia. As a result, it is very seldom that a victim of this inherited trait lives to reproduce. It was therefore surprising to discover that the gene responsible for the occurrence of the hemoglobin involved, hemoglobin S, was found in relatively high frequencies in some populations. In parts of West Africa, this gene was calculated to occur in frequencies as high as 30 percent. In addition, another abnormal hemoglobin, designated hemoglobin C, occurred in frequencies approaching 20 percent in other parts of West Africa. While not as lethal as hemoglobin S, hemoglobin C also reduced the fitness of its possessors through serious anemic conditions and its relatively common occurrence was also difficult to explain.

When information accumulated concerning these conditions, it was found that in both hemoglobin S and hemoglobin C anemias, there were two degrees of expression of the condition. In some of the affected individuals, the anemia progressed to a fatal condition early in life while in a larger segment of the population, it occurred only on certain occasions and was less often fatal. It was subsequently determined that the fatal anemia was associated with the inheritance of the hemoglobin S (or, in other areas, C) gene from both parents, while in the less lethal cases the individuals possessed the mutant gene in only single dose. This difference between homozygotes and heterozygotes was seen to occur because heterozygotes possessed normal hemoglobin (hemoglobin A) in addition to hemoglobin S. They were therefore able to sustain their metabolic oxygen requirements in all but the most strenuous circumstances. But their hemoglobin S-bearing red blood cells could be induced to sickle when the oxygen concentration of their blood fell to a sufficiently low value. This logical explanation of the difference between homozygotes and heterozygotes shed no light on the mechanism allowing the gene, which had to be considered deleterious in either single or double dose, to remain in the population in the face of strong selective pressure.

The explanation of this seeming paradox came when it was discovered that those individuals possessing both hemoglobin S and hemoglobin A suffered a lower mortality rate when infected with malaria than those individuals of the normal homozygote hemoglobin A genotype.[4] Although not fully understood, it appears

4. A. C. Allison, "Genetic Factors in Resistance to Malaria," *Annals of the New York Academy of Science* 9 (1961): 710-729; idem, "Malaria and G-6PD Deficiency," *Nature* 197 (1963): 609, idem, "Protection Afforded by Sickle-Cell Trait Against Subtertian Malarial Infection," *British Medical Journal* 1 (1964): 290-294; idem, "Population Genetics of Abnormal Hemoglobins and Glucose-6-Phosphate Dehydrogenase Deficiency," in *Abnormal Hemoglobins in Africa*, ed. G.H.P. Jonxis (Oxford: Blackwell Scientific Publications, 1965), pp. 365-391.

that the mechanism of this improved fitness in the presence of malaria involves the prevention of successful reproduction of the malarial parasite because of the sickling of the cell harboring it. It also appears that the parasite has some difficulty metabolizing hemoglobins S and C and therefore does not achieve the capacity to reproduce as quickly as would be the case if it had taken up residence in a red blood cell possessing only hemoglobin A. All red blood cells possess the same genetic information, therefore all of the red cells of an A/S heterozygote possess both hemoglobin A and S. Consequently, malarial parasites will not be able to select cells possessing their ideal environments and, while some may survive, their reproductive rate will be low and the malarial symptoms of the host comparatively mild.

In order to persist in a population a gene as lethal in the homozygote as that for hemoglobin S must confer a significant advantage on its heterozygote possessors. They must, in fact, be more fit than the normal homozygotes. This appears to be the case in areas of heavy malarial infestation, the heterozygote advantage being great enough to allow both hemoglobin S and hemoglobin C to occur in high frequencies. It is thought that these frequencies are the result of a balance of selective forces in which the superior fitness of the heterozygote compensates the population for the cost of losing the individuals homozygous for hemoglobin S each generation. The equilibrium thought to prevail in such circumstances is associated with what has been termed a balanced polymorphism, a condition which can only persist in the presence of heterozygote superiority. How might such a balance have been obtained? It is in this area of inquiry where the abundance of information concerning sickle-cell anemia makes this trait one of the best documented of all examples of adaptation in any species and therefore of interest to all biologists.

Human hemoglobin is a four-chain molecule. Normal hemoglobin A is made up of two alpha chains and two beta chains, symbolized as HbA = $\alpha^A_2 \beta^A_2$. The protein chains are synthesized *in situ* on the red blood cell. Alpha chains are synthesized at specific sites on the cell and beta chains at others. Single chains are synthesized, released, and pair in a manner which may be symbolized in the following manner: $2\alpha^A \rightarrow \alpha^A_2$ and $2\beta^A \rightarrow \beta^A_2$, a process called dimerization. The alpha and beta chains then join to form the full molecule, a reaction which may be symbolized in the following manner: $\alpha^A + \beta^A_2 \rightarrow \alpha^A_2 \beta^A_2$. Protein synthesis takes place early in the life of the red blood cell, while it still possesses a nucleus. During the rest of its life, the period during which its function is primarily that of a transport vehicle for oxygen and carbon dioxide, the red blood cell is enucleate and can therefore synthesize no new proteins. In the earliest stages of protein synthesis of each young red blood cell, it produces a hemoglobin slightly different from that characterizing the bulk of its hemoglobin compliment. This hemoglobin, called "fetal hemoglobin" has the molecular configuration $\alpha^A_2 \gamma^F_2$, indicating that it possesses gamma chains where beta chains will predominate later. Human fetuses possess much higher percentages of fetal hemoglobin than do adults, hence the name, but a small amount is always present even in adults and is referred to as "Kunkle's fraction."

As a result of several decades of sophisticated biochemical analysis, using a variety of techniques, a great deal has been learned concerning the structure of hemoglobin. In fact hemoglobin is one of the best-known proteins, having been studied in a wide variety of organisms to the extent that taxonomic affinities of a number of species have been verified through comparison of the amino acid sequences of their hemoglobins. The precise sequence of amino acids in the alpha chains and beta chains of human hemoglobin were among the first to be determined. The variations in sequence making hemoglobin S and hemoglobin C different from hemoglobin A have also been identified, as have a number of other, less frequent hemoglobin variants. From these studies we know that normal hemoglobin A possesses a total of 574 amino acids, 141 in each alpha chain and 146 in each beta chain. Since they are produced by dimerization, the alpha chains are identical to each other as are the beta chains. Detailed analysis of the amino acid sequences of the chains of hemoglobin A and comparison with those of hemoglobins S and C has shown that the alpha chains are identical in all three varieties. The beta chains are different, however, and this difference is what makes the behavior of the red blood cells possessing them differ in the manner discussed above. Considering the magnitude of the effect of possessing one or the other hemoglobin, the differences in their molecular structure turn out to be comparatively small, being limited to the substitution of a single amino acid at the sixth position in the beta chain of both hemoglobin S and hemoglobin C. The nature of the substitution is shown in Table 4. It can be seen in Table 4 that whereas glutamic acid occurs in the sixth position in hemoglobin A, that position is occupied by valine in hemoglobin S and lysine in hemoglobin C. If we pursue the inquiry a step farther to ascertain the genetic implications of these differences, we see that not only are the amino acid sequences different to a minimal extent, but the magnitude of the probable variation in the DNA sequence giving rise to it is minimal as well.

TABLE 4

CHANGES IN AMINO ACID SEQUENCE IN THE BETA CHAINS OF HEMOGLOBINS S AND C

Amino Acid Number	*HbA*	*HbS*	*HbC*
1	Valine	Valine	Valine
2	Histidine	Histidine	Histidine
3	Leucine	Leucine	Leucine
4	Threonine	Threonine	Threonine
5	Proline	Proline	Proline
6	Glutamic Acid	Valine*	Lysine*
7	Glutamic Acid	Glutamic Acid	Glutamic Acid
8	Lysine	Lysine	Lysine

TABLE 5

ALTERATIONS IN THE NUCLEIC ACID SEQUENCES ASSOCIATED WITH CHANGES IN THE SIXTH AMINO ACID OF THE BETA CHAINS IN HEMOGLOBINS S AND C

Protein	*DNA Sequence*
Hemoglobin S	C - A - T
	↑
Hemoglobin A	C - T - T
	↓
Hemoglobin C	T - T - T

This can be seen in Table 5. So we can see that the slightest perturbation in DNA sequence, a "point mutation," may have far-reaching effects on the fitness of its possessor. Beyond that, we can add that when a number of other hemoglobin variants is studied in a similar manner there is a preponderance of alterations in the sequence of amino acids making up the beta chains as opposed to those making up the alpha chains. Many variations seem to cluster at or near the sixth position, the site of substitution in the sickling hemoglobins. This raises the possibility that at this site mutation, for whatever reason, is more frequent or more viable than at other positions in the amino acid sequence. Maybe mutations occur here relatively frequently now and have so occurred in the past, a possibility to be discussed further on.

The greater frequency of viable mutations in the beta chain may be associated with the necessity of alpha chains having to interact with gamma chains in fetal hemoglobin and being therefore permitted less variation in a functional red blood cell. It has been observed that fetal hemoglobin has a greater oxygen-carrying capacity than normal hemoglobin but that immature cells possessing it in large quantities are more fragile than mature, enucleate red blood cells. Any variation which disrupted the oxygen transport system of the fetus and newborn would be severely damaging and would probably not persist in the population even if such mutations were relatively frequent. Therefore it is perhaps not surprising that variants in the beta chain are more commonly found than those in the alpha chain.

It is believed that a genetic equilibrium has been maintained in a number of African populations through the superior resistance to malaria possessed by A/S or A/C heterozygotes. It has been argued that since possessors of hemoglobin C experience less reduction in fitness than possessors of hemoglobin S, hemoglobin C would eventually displace hemoglobin S and become the most common form of hemoglobin in areas of heavy malarial infestation. The validity of this argument will probably never be fully tested, since malaria is being treated in most parts of the world and the forces maintaining the equilibrium are no longer in operation to any significant extent in most areas. Many carriers of these hemoglobin variants have lived in nonmalarial areas for their entire lives and may, in some cases represent the ninth or tenth generation since removal from the malarial environment. When a population is removed from the forces which maintain a genetic equilibrium of the sort described, we would expect the frequency of the deleterious gene to fall, eventually to a level which is just maintained by the mutation rate. There is some indication that this might be happening in populations of African ancestry living in nonmalarial areas as well as those residing in parts of Africa where prevention and treatment of malaria has been effective. It will be instructive to consider the evidence of equilibrium for hemoglobin S and hemoglobin A before turning to discussion of some of the ecological factors associated with this form of human adaptation.

We may approach the problem of calculating equilibrium levels in a balanced polymorphism by first estimating the fitness of the various genotypes or, conversely, through estimation of the selective disadvantage each genotype experiences. Since we are concerned with the intensity of selection at the locus in question favoring one or the other allele, we will also need to have some estimate of the initial allelic frequencies. With these estimates, it is possible to calculate the point at which equilibrium will occur and to predict how long it will take to reduce the frequency of a deleterious allele to a level maintained by the mutation rate. To do this we proceed in the following manner:

Genotype	A/A (p/p)	A/S (2pq)	S/S (q/q)
Phenotype	Normal	Sickle-Trait	Anemic (lethal)
Fitness	$1-S_1$	1	$1-S_2$

Where S_1 = selection against normal hemoglobin by loss of Hb A homozygotes to malaria; S_2 = selection against sickle-cell anemic individuals, which may be considered for present purposes as being 100 percent lethality.

A. P. Allison collected gene frequency data in African populations and found that when observed frequencies of adult genotypes were compared with those expected, real discrepancies arose indicating differential fitness among genotypes, heterozygotes being most fit. His findings were as follows:

	Expected	*Observed*	*Fitness*
Genotype A/A	0.6529	0.616	0.943
Genotype A/S	0.3103	0.384	1.238
Genotype S/S	0.0367	0	0

Based on observed gene frequencies:

$$A = p = .808$$
$$S = q = .192$$

Therefore, if the relative fitness of A/S heterozygotes is considered to be 1, since they reveal the greatest survival rate of the three genotypes in question, the relative fitness of the normal homozygote A/A is 0.943/1.238 = .76. The value of S_1, the selection coefficient for the normal genotype S is then $1 - .76 = .24$ and, since we see no S/S adults $1 - S_2 = 0$ and $S_2 = 1.00$.

Once we have determined these values, we are able to predict the equilibrium gene frequencies for this locus. This is possible because the word "equilibrium" tells us that all of the selection taking place will produce no net change in gene frequencies. Since we have restricted our consideration to two alleles at a single locus, we are stipulating that all of the individuals in the population will be either A/A, A/S, or S/S, i.e., in terms of gene frequencies $p + q = 1.00$. Viewed in a slightly different way, $p = 1 - q$ and $q = 1 - p$. We have stipulated equilibrium, which means that the selection against A balances the selection against S, i.e., $S_1p = S_2q$. Putting all of this together, we are now able to say that:

$S_1 (1 - q) = S_2q$. Pursuing the algebra just a bit farther, we can translate this into: $S_1 - S_1q = S_2q$, which can then be stated as: $S_1 = S_1q + S_2q$. Then: $S_1 = q (S_1 + S_2)$ and, finally, $q = \frac{S_1}{S_1 + S_2}$. Similar calculations will show us that the frequency of the normal allele, $p = \frac{S_2}{S_1 + S_2}$.

From this, it emerges that all we need to determine the equilibrium gene frequencies in a balanced polymorphism (if all of our assumptions are satisfied) is a good estimate of the selection coefficients. In the present example, the frequency of the sickle-trait heterozygotes will be 2 (0.1935 x 0.8065) = 0.3121.

Recent work in West Africa has produced estimates of a heterozygote frequency on the order of 0.22, which would not, according to the above estimate, reflect the expected equilibrium level and may be the result of reduced mortality of normal homozygotes as a result of effective medical intervention.

If we turn our attention to populations living in areas outside of the malarial zone, such as is represented by populations of black African descent living in the United States, and if we estimate the level of admixture with non-African populations at about 20 percent,[5] it is possible to test the assumption that an equilibrium existed at one time on the African continent.

Recent estimates of the frequency of heterozygote A/S genotypes among black Americans yield a figure of about 9 percent. Since this population has not been exposed to heavy selection by malaria for many generations we might fairly assume that the low frequency of hemoglobin S reflects a breakdown of the balanced polymorphism and its replacement by a transient polymorphism. The greatest fitness in the absence of malaria is possessed by the A/A homozygote. As a result, the new situation may be represented as follows:

Genotype	A/A	A/S	S/S
Phenotype	Normal	Normal	Anemic (lethal)
Fitness	1	1	0
	$(S_1 = 0)$		$(S_2 = 1)$

Since the sickle-cell anemic individual (S/S homozygote) is still unlikely to reproduce, the S gene is maintained in the population only by virtue of the viability of the heterozygote. Therefore, S/S homozygotes will continue to be produced, but in steadily decreasing frequencies after the heterozygotes cease to be the fittest genotypes. Homozygote recessives will only be born to parents who are both heterozygotes. The frequency of heterozygotes in the reproductive population, excluding the homozygote sicklers, is: $\frac{2pq}{p^2 + 2pq}$ which may be simplified by the expedient of dividing through by p to produce: $\frac{2q}{p + 2q}$. Substituting $(1 - q)$ for p we get: $\frac{2q}{1 + q}$. Consequently, the probability of a heterozygote marrying a heterozygote may be calculated as $\frac{2q}{1+q} \times \frac{2q}{1+q} = \frac{4q^2}{(1 + q)^2}$ When A/S heterozygotes mate, one quarter of their offspring may be expected to be S/S homozygotes, so that we may calculate the frequency of S/S genotypes one generation after the breakdown of a balanced polymorphism to be:

$\frac{1}{4}\left(\frac{4q^2}{(1 + q)^2}\right) = \frac{q^2}{(1 + q)^2} = \left(\frac{q}{1 + q}\right)^2$. From this we can also see that the frequency of the S allele is $\frac{q}{1 + q}$. We will state this equality as $q_1 = \frac{q_0}{1 + q_0}$.

In the next generation, the same mechanism functions to reduce S/S homozygote frequencies, so we simply express the frequency of the S allele in the second generation as $q_2 = \frac{q_1}{1 + q_1}$. Since $q_1 = \frac{q_0}{1 + q_0}$, calculation of q_2 may be done in the following manner:

$$q_2 = \left(\frac{\frac{q_0}{1 + q_0}}{1 + \frac{q_0}{1 + q_0}}\right) \text{ which reduces to}$$

$$\frac{\frac{q_0}{1 + q_0}}{\frac{1 + q_0 + q_0}{1 + q_0}} = \frac{q_0}{1 + q_0} \times \frac{1 + q_0}{1 + 2q_0} = \frac{q_0}{1 + 2q_0}$$

5. T. E. Reed, "Caucasian Geneo in American Negroes," *Science* 165 (1969):762-768.

And, it is possible to state, in more general terms: $q_n = \frac{q_0}{1 + nq_0}$, where n is the number of generations since the fitness of the homozygote A/A and that of the heterozygote A/S genotypes have become identical. If we know the equilibrium gene frequency (q_0), it is possible to calculate the gene frequency for any number of generations after the loss of the heterozygote advantage. Also, if we know the equilibrium gene frequency, which we determined earlier by using the selection coefficients, and the present gene frequency, it is possible to calculate an approximation of the number of generations since the equilibrium broke down.

Starting with $q_n = \frac{q_0}{1 + nq_0}$, we rearrange the equation to get $q_n + nq_0(q_n) = q_0$; and $n\,q_0(q_n) = q_0 - q_n$;

$nq_0 = \frac{q_0 - q_n}{q_n}$; and, finally, $n = \frac{q_0 - q_n}{q_0\,q_n}$.

Applying the frequencies derived earlier, we calculate:

$$n = \frac{.1935 - .050}{(.1935)\ (.050)} = \sim 15.$$

Assuming 20 percent migration of non-African genes,[6] the number of generations may be reduced to thirteen. The usual approximation of generation time is twenty-five years, which would give a figure of 325 years as an estimate of the length of time since the ancestors of black Americans were in genetic equilibrium for hemoglobins A and S. Since many assumptions concerning equilibrium conditions are difficult to test, the above can only be regarded as an approximation of reality with the possibility of a fairly wide margin of error. However, the plausibility of the hypothesis of genetic equilibrium is enhanced by the reasonable estimate of generation time which emerges when we apply the estimates of gene frequencies which are available. Also, the method of making the estimate is one which may be applied in other contexts and is therefore of value to students of adaptation in assessing the probability of superior fitness being possessed by heterozygotes.

Although hemoglobin S and hemoglobin C are traits which occur most often among black Africans, they occur in some other populations as well, but in very low frequencies. There has been considerable interest in attempts to reconstruct the manner in which the allele for hemoglobin S spread in African populations. A reasonable and fairly widely accepted reconstruction of these involves the change in ecological relationships of human populations residing in Tropical West Africa when the so-called "Malayan Africultural Complex" was introduced to the area.[7]

It is thought that preceding the adoption of the crops making up this complex, population densities were light and people were more mobile than is now the case. The Anopheles mosquito population was also light, since the conditions necessary to support large numbers of them and to allow them to reproduce in great numbers were not prevalent. It is also probable that the parasite responsible for the symptoms of malaria, the protozoon Plasmodium falciparum was also present. However, as long as the forest remained relatively undisturbed, the forest floor absorbed most of the water received through frequent rainfall and standing water of the sort required

6. Ibid.

7. Frank B. Livingstone, "Anthropological Implications of Sickle Cell Gene Distribution in Africa," *American Anthropologist* 60 (1958):533-557.

for mosquito breeding was comparatively scarce. Consequently, the vector responsible for transporting the protozoon parasites to their mammalian hosts was present in modest numbers. The small number of humans in any area at any one time kept the probability of a human encountering a parasite-bearing mosquito low. When infection did occur, as it must have from time to time, the probability of a vector carrying the parasite from the infected person to another was also low. Under such conditions, all of the ingredients for endemic malaria, people, mosquitoes, and parasites, were present but the frequency of the disease was low. The intensity of selection against normal A/A homozygotes would have been low as well, although there is no reason to think that A/S heterozygotes did not occur and possess superior fitness when infected even then. The above-mentioned observation of high mutability in the area of the beta chain altered in hemoglobin S makes it likely that individuals possessing the A/S genotype appeared from time to time. Their low frequency would have kept the likelihood of their mating with each other low and losses of S/S homozygotes would have consequently been low as well. It would be likely that the greatest overall fitness would have been possessed by the A/A genotypes during this period. The changes in the landscape following the shift to sedentary agriculture would soon change that, however.

When certain crops including manioc and plantain began to be exploited in tropical West Africa, it became possible to support many more people on a given amount of land. In order to take advantage of new agricultural crops and techniques, it was necessary to clear parts of the forest and to stay in the vicinity of the growing crops to tend, protect, and harvest them. Clearing parts of the forest and breaking the soil created numerous areas where standing water could accumulate. The already-present mosquitoes, including the parasite-bearing Anopheles began to appear in larger numbers and were more often successful in finding human hosts. As both the human and mosquito populations grew, the likelihood of infection increased until it became likely that most humans in the vicinity of one of these early agricultural communities would be exposed to malaria early in life. At that point, the superior fitness of the A/S heterozygotes would lead to the increase in frequency of the S allele until it reached an equilibrium of the sort previously calculated. If this is an accurate reconstruction of events, the resemblances to the sequence of events leading to alterations in phenotype frequencies in other species mentioned at the beginning of the chapter are clear. A mutation arises, confers no advantage on its possessors and consequently remains in low frequency in the population until some element of the environment is sufficiently altered to create a new demand which is, in some way, better met by possessors of the mutant allele. Thereafter, gene frequencies shift to reflect the new levels of fitness and may lead to fixation or some form of balanced or transient polymorphism. The sequence appears similar whether the affected populations are made up of humans, peppered moths, or bacteria. This, of course, is the kind of adaptation Darwin had in mind as well, but was not equipped with a well-developed genetic theory to explore in the same manner available to contemporary students of evolution.

Interestingly, adaptation to malaria appears to have taken a somewhat different course in some other human populations. In European populations of the Mediterranean region, in the Middle East, and in parts of India, a condition called thalassemia occurs in variable frequencies. It also appears to be differently manifested in homozygotes than in heterozygotes. The homozygote condition, called thalassemia major and, sometimes, Cooley's anemia, leads to a serious form of anemia which is often fatal. The heterozygotes may be detected through symptoms of anemia less serious than that of homozygotes, a condition referred to as thalassemia minor.

Underlying the condition is a lowered capacity to incorporate the amino acid leucine in the polypeptide chains making up both chains of hemoglobin. This leads to faulty hemoglobin synthesis. The synthesis of the beta chain appears most severely impaired but it appears that alpha chains cannot be released from their sites of synthesis and complex to form a complete hemoglobin molecule. As a result, chains are not released, leading to a lack of incorporation of heme groups. An excess of unincorporated heme groups, leads to a shutdown of heme synthesis, presumably as a negative feedback response.

Although leucine incorporation is impaired, the amino acid isoleucine, which is found in fetal hemoglobin but not in hemoglobin A, is incorporated as well by thalassemics as by normal individuals. All red blood cells appear to produce some fetal hemoglobin, even in adults. This is usually a small fraction synthesized early in the life of the cell. As the cell matures, synthesis of gamma chains ceases and is replaced by beta chain synthesis. In many forms of severe demand on the blood-forming tissues, the maturation of red blood cells is altered so that many enter the circulatory system while still producing fetal hemoglobin. It appears that an inherited defect in beta chain synthesis leads to such a circumstance with the result that a larger fraction of the hemoglobin is of the fetal variety. In thalassemia major, as much as 90 percent of the total hemoglobin complement may be fetal hemoglobin. Since this is an indication of a large number of immature, fragile red blood cells, the blood forming tissues of these individuals are severely overtaxed. The result is a form of stress which leads to early mortality. The anemia of the heterozygote is less severe, but is also characterized by a higher percentage of fetal hemoglobin. Since red blood cells appear in the circulation in a more vulnerable state in any form of anemia, they are, even in the thalassemia minor phenotype, less capable of harboring the malarial parasite sufficiently long to allow its successful reproduction. Destruction of such cells and their parasites would keep the level of infection relatively low and allow survival at the cost of some reduction in vitality for such heterozygotes. The selective advantage does not appear as clear-cut as in the sickle-cell heterozygotes, but there is statistical evidence that the frequency of the allele responsible for thalassemia rises and falls in direct proportion to the probability of exposure to tertian malaria in the regions of its occurrence.

Although these two conditions, sickle-cell anemia and thalassemia, are apparently associated with the occurrence of malaria and both involve the important protein hemoglobin, they are fundamentally different responses. The sickle-cell trait involves a real change, presumably through a mutation, in the structure of the protein by substitution of one amino acid for another. In thalassemia, the change is in the ability of the protein synthesis mechanism to function. This is possibly due to a defect in the synthesis of the messenger RNA which must leave the nucleus and induce the aggregation of the proper sequence of amino acids on the polyribosomes.[8] That both conditions appear to lend increased fitness to heterozygotes exposed to malaria is testimony to the variety of pathways the species may exploit to adapt to a sufficiently strong selective force. These are both genetic responses but one involves a structural protein while the other involves, in all likelihood, an enzyme concerned with an aspect of protein synthesis. Another metabolic defect which is inherited and appears to be associated with resistance to malaria is a condition called G-6PD deficiency. The way in which the defective enzyme synthesis invoked in this condition affords increased malarial resistance remains a matter for speculation, but statistical analysis of its distribution has led to its being included among the traits which enhance the survival potential of their possessors in the presence of malaria. There may well be others.

The examples discussed above give some idea of how a population may exploit a new genetic trait to cope with a strong environmental stressor. Interaction of human populations with the malaria parasite have provided biologists with important examples of how adaptation works. In many other cases, the knowledge of the mechanisms involved is not nearly as complete, and many points of the connection between cause and effect must be inferred. For that reason, some of the most impressive examples of natural selection operating on human populations can only be dealt with superficially and do not contribute to the understanding of the adaptive process to the extent that the foregoing examples do. Nevertheless, any discussions of human adaptability would be deficient without some mention of the intense selective pressure exerted by a number of contagious diseases occurring in epidemic proportions.

THE IMMUNE SYSTEM AND ADAPTATION

When human populations became more sedentary and began to live closer together for extended periods of time, they became much more vulnerable to the ravages of contagious diseases. It is probable that as human populations have become increasingly sedentary, the capacity of individuals to survive has become dependent to an increasing degree on the effective function of the immune response. While life became easier in many other respects, disease resistance has become the frequent determinant of who would live to complete reproductive life and who would not. If we compare contemporary human populations we see that resistance to disease varies greatly from place to place. On occasions when diseases with relatively benign effects in one population have been introduced to populations which had not previously encountered them, the result has often been disasterous. The decimation of American Indian populations by measles,

8. Many aspects of current research in the etiology of thalassemia may be found in the papers included in the *Third Conference on Cooley's Anemia*, ed. Edward Zaino, *Annals of the New York Academy of Science* 232 (1974).

chicken pox, and small pox is a case in point. Even today, South American Indian populations appear to be susceptible to severe effects when infected by a wide variety of upper respiratory diseases that are only mildly debilitating to Europeans and North Americans. This is true despite the fact that many of the same individuals who are severely affected possess extremely high concentrations of antibodies in their circulation, presumably the wrong ones for coping with influenza viruses carried by the people who visit them.[9]

In many cases, it is reasonable to argue that the disease organism and the host accommodate to each other in a form of what might be regarded as a mutual adaptation. This appears to be the case with many human populations and the measles virus. It is, after all, not beneficial to the parasite if its activities kill the host and thereby destroy its environment. If the parasite can adjust its demands to allow the host organism to survive, it improves its chances to reproduce and spread. This capacity, conferring a selective advantage, would be possessed by an increasing proportion of the organisms involved over time. Since disease organisms have a very short generation time compared to their human hosts, it would be expected that they would evolve more rapidly. It would be expected that mutual interests of the disease organisms and the hosts will converge at a point which does not necessarily involve a total destruction of the disease organisms by the immune response. We, along with most other species, harbor a number of organisms which contribute to our well-being, as do the lactobacilli of the human intestinal tract. The advantages of retaining such parasites favor some degree of immune tolerance for organisms which are present with benign effects. However, since the mutual adaptation which has taken place in one area has not necessarily occurred in another, exposure to new organisms can be traumatic, even fatal. In addition, the immune system functions in two distinct phases, the first of which requires the "learning" of a specific synthetic pathway by certain cells when they recognize a new antigenic substance. This process is slow, taking four or five days. Subsequent exposure to the same antigen evokes a strong and rapid response which usually is capable of dealing with the infecting organism. Often, first exposure to antigens occurs shortly after birth, when a baby is protected by passive immunity conferred by the mother *in utero* and through nursing. So while protected against the virulent effects of disease organisms, the infant establishes an immune response system which is prepared to generate rapid production of antibodies in a secondary immune response when later exposure to a disease organism warrants it. In populations which have not been exposed to the organism, the primary response has not been established and, in many cases, the disease organism has proliferated to a damaging extent before an effective antibody complement can be produced. The results of such a delay can be, and often are, of fatal consequence. We do not know of any genetically transmitted immunities to specific diseases, but there are probably individual variations in the capacity to produce antibodies and there may be population differences as well.

During a period lasting from the middle of the fourteenth century to the middle of

9. James V. Neel, "Lessons from a 'Primitive' People," *Science* 170 (1970): 815-822.

the eighteenth century, Europe experienced a series of epidemics which severely decimated its population. These epidemics, collectively called "The Plagues," involved a number of diseases but the organism which appears to have been the most virulent was Pasteurella pestis. This species of bacteria affects humans in several distinct ways, all of which have a high probability of ending in death. The diseases have been called Bubonic, Pneumonic, and Septicemic Plague depending upon whether the symptoms primarily involved the lymphatics, pulmonary system, or blood. The course of the disease was usually rapid and fatal in a high percentage of victims. Even when a victim survived the ravages of plague, his weakened state frequently made him easy prey to secondary infections, so that pneumonia, influenza, scarlet fever, and smallpox also took their toll during this period.[10]

It is possible that similar epidemics occurred earlier, particularly during the later years of the Roman Empire, but never before or since has the toll of human life to disease been so great as during the approximately 400 years of the great plagues. There has been no fully adequate answer to the question of why the plagues were so widespread and severe during this relatively limited period in human history, but certain facts give some indications of the reasons. For one thing, humans had come to live in relatively congested urban communities in much of Europe during this period. High population density was achieved without the benefit of sanitation and public health measures which are now recognized as essential. Therefore, garbage and sewage accumulated in residential areas, and rats and mice were present in large numbers. Rats carry fleas and fleas are effective vectors for Pasteurella pestis. That much of the contagion resulted from contact with bacteria-bearing fleas seems to be quite certain. The epidemics followed seasonal cycles, with peaks during the winter months when people tended to cluster indoors, followed by periods of relatively light mortalities during the summer. The mortalities were highest in the large urban centers. Most of the major cities of Europe felt the effects of the epidemics at some point during the four-century period, many of them repeatedly. Some cities lost over half their populations through death and emigration. The economies of the smaller agricultural towns surrounding the major cities also suffered and these communities experienced a loss of population through emigration. All in all, the continent of Europe was subjected to a series of demographic changes which probably exceeded in variety and magnitude any that have occurred before or since. High mortalities, migrations, breakdown of community life, famines, and many other forms of dislocation were faced by large numbers of people until, in the eighteenth century, for reasons not yet understood, the plagues ended.

It is likely that the population of Europe in 1750 was quite different from that living in 1350 in a number of respects. One thing is quite certain; the immune systems of the survivors of sixteen or more generations of severe selection in the face of an unprecedented number of

10. A readable and highly informative account of the plague epidemic in Europe is found in: W. L. Langer, "The Black Death," *Scientific American* 210 (1964): 114-121.

pathogenic circumstances were effective. So it is probable that the immune systems of European populations are still, to some extent, adapted to infectious diseases. Some of the adaptation may be of the sort in which pathogen and host tolerate each other. Possibly more important is the capacity of the immune system to produce antibodies rapidly when exposed to possibly dangerous antigens. This latter capacity might be effected by the continued presence of some harmful organisms stimulating a primary immune response early in life. If later exposure to the organism is sufficiently severe, the secondary immune response comes into play preventing serious damage to the host. If such a system is widespread in the human population, it provides a means of understanding how some populations tolerate certain diseases but are susceptible to others. This is an adaptation made up of both genetic and extragenetic components, one which is of great importance to the survival of the species and benefits greatly from the flexibility inherent in the possession of genetically determined traits capable of variable expression. It is probable that resistance to Pasteurella pestis had to be "learned" by the immune system of Europeans and initial exposure caught it off guard. It is probable that the organism was first imported from the East. Apparently, ships landing in Italy had picked up flea-bearing rats at Black Sea ports and the first phase of the plagues took place in Italian port cities, subsequently spreading north. Perhaps the populations from which the organisms came were less susceptible to its effects because of less severe crowding and the European continent presented Pasteurella pestis with its first opportunity to multiply in great numbers. Other diseases including Typhus and Cholera have attained epidemic proportions in other parts of the world, but the fact of their continued occurrence indicates that no genuinely effective adaptation to them has yet become widespread.

Perhaps other examples of genetic adaptation to environmental stress in human populations exist, but when examined closely, many human adaptations prove to be of the sort categorized in chapter one as acclimatizations. It is this form of very useful device for adjusting to the demands of the environment that will be the subject of the remaining discussion.

For Further Reading

Crow, James F. *Genetics Notes.* 7th ed. Minneapolis: Burgess, 1973. A useful collection of genetic analytic techniques. One of the clearest, most concisely written sources available.

Dobzhansky, Th. *The Genetics of the Evolutionary Process.* New York: Columbia University Press, 1970. A very useful reference for the student with some mathematical and theoretical background. Even the beginner will get some ideas from reading this compendium, but it will take considerable effort.

Johnston, Francis E. *Microevolution in Human Populations.* Englewood Cliffs, N.J.: Prentice-Hall, 1973. A good place for the beginning student to learn techniques for the measurement of gene frequency changes in human populations.

Lerner, I. M. *Genetics, Evolution and Society.* San Francisco: Freeman Press, 1968. A useful synthesis of general genetics and bits and pieces of human biology assembled for the beginning student.

Otten, Charlotte M. "On Pestilence, Diet, Natural Selection and the Distribution of Microbial and Human Blood Group Antigens and Antibodies." *Current Anthropology* 8/3 (1967):209-226. A penetrating review of several theories concerning the interaction of disease, self-recognition by the immune system and human blood group antigens during the great plagues.

Wilson, Edward O. and Bossert, William H. *A Primer of Population Biology.* Stamford, Connecticut: Sinauer Associates, 1971. Introduction to the mathematical treatment of evolution. Indispensable to the student contemplating further work in quantitative ecology.

BIBLIOGRAPHY

Appleman, Philip, ed. 1970. *Darwin*. New York: W. W. Norton Co.

Buettner-Janusch, John. 1966. *Origins of Man.* New York: John C. Wiley and Sons Co.

Morris, Laura Newell, ed. 1971. *Human Populations, Genetic Variation and Evolution.* San Francisco: Chandler Publishing Co.

4 The Human Capacity to Adjust

From the standpoint of the individual, the evolution of the species is, in many respects, a closed book. The equipment for survival has been inherited to be used for maximum benefit. The central concern of the individual organism is survival—survival as a unit, not as a collection of organs or organ systems, but as a functioning entity all of whose constituents are mutually dependent.

Unlike the population or the species, the organism cannot benefit from superior fitness of one of its components at the expense of another. It must, by whatever means, preserve its integrity. This is so even when self-preservation might be contrary to the long-term interest of the species. The strategy of the genes is replaced by the tactics of survival. A starving organism wastes none of its energy on reproductive activity, even if it is the last of the line and its failure to reproduce spells extinction for the species. Its final efforts will instead be directed to the perpetuation of its own existence, oblivious to the grand design of the evolutionary process. Most of the time, of course, the instinct of self-preservation of its members works to improve the chances for survival of the species. Since each individual carries the future of the species in its germinal tissue, self-interest seldom conflicts with the interest of the species. As a result, evolution has favored adaptations which permit organisms to make whatever adjustments are required to cope with changes in the environment experienced during a lifetime. The greatest benefit is gained from early adjustments since they permit survival to reproduce. So the capacity for making large adjustments in physiological and morphological characteristics is of greatest value during growth and development. Other, less drastic changes involve a smaller biological investment and may be made throughout the life of the individual. The capacity to make significant physiological adjustments decreases with increasing age, which is also in line with the interests of the species, since the preservation of individuals past the reproductive period has little value for most species. This is not entirely true for species in which the welfare of the population may be enhanced by the presence of older members who may contribute skills and knowledge which improve the survival rate of the young, maximizing reproductive success and possibly allowing the expansion of the range. The strategy employed by our species appears to favor the survival of individuals past their reproductive period. Associated with this longer nonreproductive life span, we are well endowed with a capacity for a number of acclimatizations, acclimations, and other adjustments throughout life. The dimensions of some of these adjustments may be truly impressive as the subsequent sections will reveal. The range and variety of these nongenetic mechanisms of survival provide us, as a species, with numerous opportunities, the exploitation which has had an important influence on the shaping of the species as we know it today.

ADAPTATION TO HIGH ALTITUDE

As was pointed out in chapter two, our mammalian heritage has provided us with a wide range of opportunities through improved homeostatic control mechanisms. We are equipped to maintain high activity levels, expand our range into areas where low temperatures occur, and move about freely where it is hot with the help of a cooling system importantly including the capacity to sweat. These opportunities are not without their costs, however. One major common element in all of these capabilities is that they all cost energy. Since we rely on oxidative metabolism for a large part of our energy supply, we require fuel, in the form of glucose as well as oxygen to combine with it in the biochemical pathways Kleiber has appropriately named "The Fire of Life."[1] There are several ways to obtain glucose. We are provided with biochemical pathways which may break down starches and complex sugars to simple sugars and with other pathways which allow the conversion of both proteins and fats into readily consumable carbohydrates.

For the most part, though, oxygen is oxygen and must be constantly available in sufficient quantity if we are to survive. At sea level, this seldom presents a problem to a healthy individual. Almost 21 percent of the atmosphere is made up of oxygen. Each breath we take is therefore rich in this essential gas. In fact, the percentage of oxygen is the same in the upper atmosphere as well. A problem arises at high altitude because the barometric pressure falls with increasing distance from the earth. The gases present in the sea level atmosphere will raise a column of mercury 760 mm against the force of gravity in a vacuum. At 13,000 feet (4,000 m), the pressure is under 465 mm. So, an equivalent breath will supply only about 60 percent as much oxygen at 4,000 meters. Most of us find this difference distressing and will involuntarily breathe more rapidly to inspire the amount of oxygen essential for our metabolic demands. Some persons, particularly among those past middle age, will not be able to adjust and must return to lower altitudes to survive. Many unacclimatized individuals will have difficulty making an effective long-term adjustment which would permit them to work, reproduce, and become successful residents of the high altitude environment.

Nevertheless, in two areas of the earth large populations of humans live their entire lives at altitudes in excess of 4,000 meters. These places, in the Himalayas of Asia and the Andes of South America, are

1. Max Klieber, "Respiratory Exchange and Metabolic Rate," in *Handbook of Physiology*, section III, vol. 2 (Baltimore: Williams and Wilkins, 1965), p. 927.

peopled by well-adapted populations which in their aggregate number in the millions. These people show considerable anatomical evidence of their adaptation and physiological tests show that they are in many respects different from lowlanders. Although separated by thousands of miles and with relatively little chance of genetic exchange, the Andean and Himalayan populations resemble each other in a number of the anatomical and physiological characteristics that appear to be associated with their successful survival under conditions of chronic hypoxia. There are a number of stresses besides hypoxia to deal with in a high altitude environment; cold, wind, rough country, solar radiation, and the scarcity of many kinds of food among them. There are ways to deal with all of these stressors except hypoxia behaviorly or culturally. Hypoxia is, however, ubiquitous and not amenable to correction. As a consequence, survival requires physiological adjustments which make the amount of oxygen available to the tissues similar to that provided at sea level. The range of adjustments and the number of homeostatic mechanisms they involve is truly impressive. Of course, one of the most directly concerned is the respiratory system, which for present purposes includes the oxygen-transport function of the cardiovascular system as an integral unit.

Since humans are erect bipeds, our system of breathing has been altered somewhat from the conventional mammalian pattern found in quadripeds. Changes in the positions of the organs in upright posture have led to secondary changes in connective tissue support and have been accompanied by an altered shape of the thorax. These are all part of our arboreal primate heritage. With a free forelimb we have reduced restriction of movement of the pectoral girdle. We also have a broad sternum which is firmly attached to the clavicle. Our ribs lie at an angle to the spinal column, being lower on their ventral aspects than at their point of articulation with the spinal column. The ribs are linked together by a set of intercostal muscles allowing the entire rib cage to be expanded as well as moved as a unit. As a result of this fortunate combination of anatomical characteristics, we are able to increase the internal dimensions of our thorax in two ways. First, we contract muscles in the diaphragm which cause it to press down on the abdominal viscera thereby enlarging the space in the thoracic cavity at the expense of the abdominal cavity. The enlargement of the cavity around the lungs in the normal thorax immediately lowers the surrounding gaseous pressure. This negative pressure around the lungs causes them to expand passively. This in turn creates a negative pressure in the alveoli of the lungs and in the airways leading to them. Outside air is drawn into the lungs and airways as a result of the fall in internal pressure creating a pressure gradient. Gaseous diffusion does the rest. An example of the principle at work may be seen when a limp, sealed balloon appears to inflate as the surrounding air is removed, an experiment often performed using a bell jar. Any factor which prevents the formation of a suitable pressure gradient interferes with the capacity to breathe, as when the thorax is penetrated by a foreign object and is no longer air tight. This potentially dangerous condition is called pneumothorax and can cause death through suffocation.

The increase in intrathoracic volume

created by contraction of diaphragm muscles is enhanced in humans by contractions of the anterior neck muscles pulling the thorax upward as a unit. The ribs, which were originally oriented downward from their point of articulation with the spinal column, are pulled up toward a horizontal position, thereby increasing the distance between the sternum and the spinal column by as much as 20 percent. As a result, the volume of the thoracic cavity is increased to a greater extent than simple diaphragm breathing would allow. We use this capacity to good effect during heavy exercise when oxygen demand is high as can be readily observed when athletes take deep breaths immediately before a strenuous effort. It should be noted that when full inspiration is thus performed, the cross section of the chest is much closer to a circular shape than when relaxed, the human thorax being apelike in its anterior-posterior flattening when at rest.

A normal breath exchanges about 500-600 cc of air. This is called a normal "tidal volume". Of this amount, about 360 cc reaches the alveoli, the rest fills the air tubes and is called dead space. Each breath of inspired alveolar air is mixed with about two and one-half liters of air already present in the alveoli. This might be considered a form of buffering against destructive changes occurring in the gaseous mixture inhaled, since a single breath of all but the most noxious of gases is not overwhelming and may be expelled before damage occurs.

In the alveoli the exchange of oxygen and carbon dioxide with the blood occurs across the walls of the alveolar capillaries. Concentration gradients of oxygen and carbon dioxide largely determine the direction of diffusion which occurs, so that in the lungs, alveolar oxygen crosses into the circulatory system and carbon dioxide passes into the lungs out of the circulatory system. In the tissues, the direction of flow is reversed. The amount of oxygen taken up and released is influenced by the *pH* of the blood, which in turn is regulated by a buffering system employing bicarbonate ions and carbon dioxide. Changes in serum *pH* have a more pronounced effect at the tissue level, since increases in carbon dioxide concentration and a decrease in serum *pH* facilitate the release of oxygen by hemoglobin, thereby improving the "unloading" of oxygen to the tissues during times when the oxygen saturation of the blood is low. The shift in oxygen release accompanying greater acidity of the blood (called the "Bohr Shift") acts as a buffer protecting the tissues from oxygen deprivation. The shift toward greater acidity of the blood is caused by buildup of carbon dioxide and resultant carbonic acid in the blood as a result of cellular metabolism.[2]

The central nervous system participates in the regulation of concentrations of oxygen and carbon dioxide. The respiratory control center is located in the hypothalamus. It receives information from a number of organs and tissues, including the aortic and carotid bodies which are sensing devices detecting increased carbon dioxide concentration and decreased oxygen in the blood. From the hypothalamus, impulses flow out through the Vagus nerve (the tenth cranial nerve), eliciting an increased or decreased respiratory rate. The participation of the

2. For a full discussion of the buffer systems see: Arthur C. Guyton, *Basic Human Physiology: Normal Functions and Mechanisms of Disease* (Philadelphia: Saunders, 1971), pp. 305-317.

sympathetic nervous system also involves the adrenal medulla which secretes epinephrine calling forth appropriate responses by the heart, kidney, and gut. All of these organs have a part to play in response to any significant decrease in the availability of oxygen upon which the whole organism is so dependent.

Respiration Under Conditions of Hypoxia

The response to altitude stress is an extension of the normal respiratory function with the following list of adjustments.

1. Rapid breathing makes more oxygen available to the lungs.
2. Dead space is reduced, increasing the proportion of inspired air made available for extraction of oxygen.
3. The lungs (alveoli) expand the area for oxygen and carbon dioxide exchange.
4. The oxygen transport system alters its function to permit maintenance of an oxygen reserve so that exercise does not create a shortage in the tissues.
5. A series of adjustments make it possible to release oxygen to the tissues at the normal concentration gradient while at the same time oxygen is picked up in the lungs at a lower than normal pressure gradient.
6. *pH* is maintained within normal physiological limits.
7. The whole series of adjustments is graded and takes place in a coordinated fashion, avoiding the trauma associated with overcompensation.

The Response of Unacclimatized Newcomers to High Altitude

A new arrival at the high altitude environment is frequently unaffected by the lowered oxygen supply, unless called upon to perform some form of physical exercise. There are some individuals who are taxed to the limit by very mild exercise while others appear to adjust readily. The usual first sign of insufficient oxygen is accelerated breathing. This can lead to hyperventilation which results in the loss of greater than normal amounts of carbon dioxide. If persistent, such loss causes a shift of the *pH* of the blood toward the alkaline, a condition of respiratory alkalosis. This and other biochemical distress signals rapidly evoke an adrenal response through the participation of the central and autonomic nervous systems. Within hours, an important aspect of the respiratory response mechanism, release of stored red blood cells into the bloodstream, takes place. This release furnishes additional hemoglobin to the blood and makes possible the maximal utilization of the oxygen present in the alveolar air. The short-term increase in the amount of circulating hemoglobin is amplified in more extended exposure by a stimulation of blood-forming tissue to increased effort. If necessary, the blood forming capacity of the body is increased. The body is capable of increasing its blood-forming tissue through growth of new red marrow. At high altitude, it appears that such expansion is limited, for the most part, to the thoracic region. In short-term exposure, the liver and spleen may be the only tissue involved. There is, in addition, a dilation of the capillaries of the lungs and increased heart rate. The kidneys participate by selective excretion of ions, a mechanism which assists in maintaining blood *pH* within normal physiological limits. The mechanism also includes neurohormonal secretions of the hypothalamus and pituitary. A common constituent of all the nervous, respiratory, and excretory functions is the control of membrane permeabilities, a direct function of ionic concentrations in the blood and interstitial fluid. The critical importance of maintaining the blood-buffer system is therefore not surprising in altitude stress.

The blood undergoes a change in viscosity as a result of the increased number of red cells. This results in increased work for the heart, especially for the right ventricle which must push more and thicker blood through an enlarged alveolar capillary bed with each stroke.

It appears that prolonged stays at high altitude result in the incorporation of these changes into the normal function of the organism. This is important, since it demonstrates that a stress can become an integral part of the environment even when first encountered in the adult state and, even then, cause no appreciable loss of viability. This sort of adaptability is not lost by the human organism throughout life, i.e., it is integral to its physiological character and is not merely a developmental feature. In fact, the human adult exposed for a period of years to high altitudes will eventually moderate the stress response in certain respects. The blood *pH* tends to return to normal and the respiration rate will return to somewhere near sea level values as the system adjusts to the maintenance of an oxygen reserve tied up in oxyhemoglobin. These adjustments may, however, take years. The morphological correlates of the physiological adjustments made by the adult are minor in nature, sometimes not observable except through serological tests. Long-term adjustment can be reflected in enhanced red marrow area, but this can only be determined if the initial values can be shown. Even then, other factors could cause identical effects, so evidence of morphological change is at best equivocal. The acclimatization can also be shown to be reversible upon return to sea level, with red cell number and serum *pH* eventually returning to normal sea level values.

THE CHARACTERISTICS OF NATIVE HIGH ALTITUDE DWELLERS

Table 6 shows some of the differences in the constituents of the blood of men living at sea level, 3,000 meters, 4,000 meters, and 5,000 meters in the Andes. These data, compiled by Carlos Monge over a number of years give some notion of the direction the adjustment to low oxygen partial pressure will take. Numerous other studies have added to the information concerning physiological changes occurring at high altitude and have shown that there remains, even in lifelong residents of high altitude environments, considerable variability. Not everyone living at high altitude exhibits all of the changes so clearly evident in Table 6 to the extent Monge's samples did. Nevertheless, the bulk of the evidence still supports the contention that the oxygen transport system responds to the demands made on it at high altitude by releasing more red blood cells into the circulation and by maintaining the increased number for extended periods, perhaps for a lifetime. Evidence collected in the Himalayas is not in full agreement with that from the Andes. The increase in the number of red blood cells in native Himalayans does not appear to be as great as the Andean subjects studied by Monge, raising the possibility that additional adaptations, perhaps at the level of cellular metabolism, may have been made by the Himalayan high altitude dwellers. Be that as it may, the serological changes of the Andean natives appearing in Table 6 are informative in several important respects. For one thing, there is no reason to think that the increase in the quantity of hemoglobin present in the increased number of red cells represents a change in the genetic code of the individuals involved from that of individuals born at sea level. This is not because of a lack of time; some of these populations may have had ancestors living at high altitudes for the last two thousand years or more. Rather, the adjustments reflected in the serological data of Table 6, are within the capacity of many, perhaps most, sea level dwellers. Even the lactic acid level associated as it is with the tendency toward a state of acidosis, as opposed to the alkalosis characterizing unacclimatized newcomers to high altitude, may be found in sea level dwellers after heavy exercise. However, the tendency toward persistant acidosis in high altitude natives does give them an advantage in unloading oxygen to the tissues through the "Bohr Shift" phenomenon mentioned earlier. We might quite legitimately view the entire complex of changes as a mere extension of the short-term capacity most sea level dwellers exhibit when acclimating to the high altitude environment. Even persistent acidosis might well be the result of prolonged anaerobic metabolism in the presence of lowered oxygen concentration. The accumulation of acidic by-products of anaerobic metabolism causes the surrounding tissue fluids to become acidic and this leads to the occurrence of the Bohr Shift, more oxygen release occurs as a result, and the requirements for oxidative metabolism are again supplied. When oxidative metabolism resumes, acids are removed through oxidation and the Bohr Shift is reversed. Altogether, the system acts as an effective servo-mechanism which functions through feedback in the form of *pH* changes, is self-correcting, and functions most effectively at a mildly acidic *pH*.

High altitude dwellers in both the Andes and the Himalayas share the characteristic of a reduced respiratory rate, which, on

TABLE 6

SOME PHYSIOLOGICAL DIFFERENCES
BETWEEN MALES FROM SEA LEVEL, AND HIGH ALTITUDE ENVIRONMENTS*

Characteristics	*Men from Sea Level*	*Natives at 3000m*	*Natives of 4000m*	*Natives of 5000m*
Barometric Pressure	750mm	518mm	480mm	446mm
Partial Pressure of Oxygen	150mm	104mm	96mm	89mm
Blood Volume (liters)	5.21	5.36	6.15	6.98
Red Cell Volume (liters)	2.34	2.79	3.36	4.29
Total Hemoglobin (gm)	788.0	905.0	1150.0	1464.0
Relative Hemoglobin (gm/100ml)	16.0	16.85	18.82	20.76
Lactic Acid (mg/100ml)	11.0	12.8	12.6	14.1
Red Blood Cells ($10^6/mm^3$)	5.14	5.65	5.67	6.14

*Adapted from Monge, C. (1960). *Aclimatacion en Los Andes* Lima: Facultad de Medicina.

the face of it, seems paradoxical. However, there are resemblances to the response to asphyxia exhibited by mammals in general involving a slower pulse and a general slowing of cellular metabolism. A pronounced expression of this form of adaptation is seen in hibernating mammals where peripheral tissues reduce their oxygen demand and thereby protect the function of the central nervous system.

The complex of serological changes involves an increase in the quantity of red blood cells proportionately greater than the increase in the total amount of blood. As Table 6 shows, the condition makes each drop of blood richer in hemoglobin content and thereby in oxygen-carrying potential. The side effect of the proportional increase of red blood cells is a "thickening" of the blood, a characteristic reflected in greater blood viscosity. Pumping a fluid of increased viscosity places a greater burden on the heart. As a result, the heart, being predominantly muscle does what any heavily exercised muscle would do —it hypertrophies. The hypertrophy is of a specific form involving the greatest amount of change in the right ventricle. If we reflect on some of the important mammalian characteristics associated with higher activity levels and greater oxygen demand on the part of the central nervous system, the reason for this particular type of cardiac hypertrophy becomes clear. The right ventricle in the mammalian four-chambered heart pumps blood to the lungs where it picks up oxygen and releases carbon dioxide before returning to the heart, to be pumped out to the rest of the body including the brain. This dual pumping system, characterized by a separate pulmonary circuit has the important advantage of preventing the oxygen-rich blood from the lungs from mixing with the oxygen-depleted blood from the tissues. Combined with the sensing devices and respiratory control mechanisms mentioned at the beginning of this section, this system allows the maintenance of a well-regulated oxygen supply to the tissues, particularly to the highly sensitive and demanding brain.

Under prolonged hypoxic stress, the heart must pump blood through lungs that are larger and have many more capillaries which allow a significant increase in the diffusing capacity of the lungs. In these larger lungs, a narrowing of the lumen of the small arteries and arterioles increases the resistance of the system while improving the exchange of respiratory gases under reduced barometric pressure. The general state of pulmonary hypertension creates an added demand on the blood vessels. The pulmonary vessels reflect the increased demand by a thickening of their walls. Along with these changes is seen a general increase in muscularization of the arteries throughout the body. Some of the characteristics of the pulmonary vessels, particularly their degree of elasticity in adults living in chronic hypoxia, causes them to resemble the tissues of fetuses and newborn infants. The retention of some fetal characteristics in the vessels of the heart and lung appears to be an economical and easily attained adjustment to the demands of an oxygen-poor environment. This pattern is found in fetuses at sea level, but is lost early under normal circumstances. It has been argued that a state of pulmonary hypertension may be closely correlated with the altitude of the place of birth with the first clear-cut evidence of its presence being found at about 3500 m. With this combination of traits we see anatomical changes that are clearly associated with physiological adjustments under prolonged exposure to the stress of low oxygen availability at high altitude. The chain of events does not end with these alterations of muscle development, which are, after all, mostly invisible to the casual observer. High altitude natives in widely separated parts of the world share certain anatomical characteristics which are clearly visible and which appear to be associated with successful acclimatization to their demanding environment. Table 7 gives some comparisons of anthropometric characteristics of sea level and high altitude populations drawn from similar racial stock.

In Table 7, there are some clearly discernible differences between the mean values presented, the most important ones being those associated with increased thoracic volume. The statures of the two high altitude populations (both all-male) are shorter. This characteristic may well be associated with a general response to stress. We infer this because populations living at high altitude resemble those found in areas of frequent famine, epidemic, and other environments of recurrent stress. In general, such populations experience slow growth, delayed maturation, and reduction or absence of the characteristic prepubescent growth spurt seen in most human populations. There will be more said on this topic in a later section.

Returning to consideration of the characteristics of the thoracic measurements of high altitude populations, certain important associations emerge. Despite the similarity of the transverse diameters of the thorax seen in the three populations compared in Table 7, there are considerable differences in their anterior-posterior diameters. The higher altitude population comes closest to approximating a circular thoracic cross section. As a result, despite their shorter stature, the high altitude natives have a greater thoracic capacity than natives of the lower altitudes. It should be noted that the difference in statures between high and low altitude natives appears to be largely associated with the possession of shorter legs by the high altitude natives. The combination of increased thoracic volume and shorter legs produces the characteristic "barrel-chested" appearance associated with dwellers at high altitudes in both Asia and South America. The means by which this chest develops under persistant hypoxic stress is a useful

TABLE 7

COMPARISON OF SOME ANTHROPOMETRIC TRAITS OF THREE SOUTH AMERICAN POPULATIONS FROM DIFFERENT ALTITUDES*

Number of Individuals	*Mean Altitude(m)*	*Average Stature (cm)*	*Anterior-Posterior Thoracic Diameter (mm)*	*Transverse Thoracic Diameter (mm)*	*Sternal Length (mm)*	*Thoracic Volume (cm^3)*
120	100	165.3	203.6	283.5	183.3	10544
124	3300	160.3	208.3	284.0	185.0	11019
53	4500	161.6	213.0	283.0	199.0	12150

*Adapted from Monge, C. (1960). *Aclimatacion en los Andes* Lima: Facultad de Medicina.

example of how growth can accommodate to the demands of the environment.

In the so-called "altitude thorax," the rib cage is permanently fixed in a position similar to that achieved by a lowland dweller when forcing a deep breath and pulling the rib cage up by contraction of the anterior neck muscles. As a result, the ribs are oriented more horizontally than those of lowlanders and the lower border of the rib cage does not taper inward toward the spinal column, but instead approximates the same diameter as the middle of the rib cage, hence the "barrel" configuration. In addition, as shown in Table 7, the length of the sternum is greater in the high altitude population. The increase in sternal length enhances the barrel-chest tendency while accommodating additional red marrow area for the production of red blood cells which are present in significantly greater numbers. The larger thorax accommodates the larger lungs which in turn possess a larger capillary network and a consequent increase in diffusing capacity. Associated with the greater diameter of the lower thorax is a greater diameter of the cross-sectional area occupied by the diaphragm. This appears to be utilized by a larger diaphragm which effectively compensates for the fixation of the rib cage in the function of increasing intrathoracic volume during inspiration. Interestingly, there is another synergism involved in the increased size of the lower ribs, allowing inclusion of more red marrow. This increases the capacity to form red blood cells, thereby amplifying the effect of the larger sternum. The net result of these increases in red marrow area is a significantly increased capacity to maintain the large complement of red blood cells characteristic of high altitude dwellers.

The human skeleton is a dynamic entity with considerable capacity to respond to the demands of the tissues around it. The apparent inconsistency between the exuberant growth of the thorax and that of the limbs under hypoxic stress reflects this capacity. From earliest infancy, the high altitude native is faced with an inescapable demand on the respiratory system. Oxygen is essential in sufficient quantities to permit continued support of a large and rapidly growing brain. As activity levels increase, so does the need for oxygen, all of which must be acquired from the thin mixture of gases characterizing the high altitude environment. In response, lung tissue grows rapidly and the associated muscles of respiration create a state of permanent tension on the elements of the thoracic skeleton. Bone has the capacity to respond to such stress with increased growth and the ribs and sternum exploit this capacity to increase their size producing the changes recognized as the altitude thorax.

It is unlikely that all individuals in a population possess an identical capacity to make these adjustments. But it is probable that individuals who are significantly deficient in this respect will survive to reproduce less often than those who are successful. This is different than saying that the altitude thorax is a genetic trait, but it does imply that whatever genetic characteristics underlie the capacity to make the necessary adjustments will be at a selective advantage under altitude stress and possibly under other forms of stress as well. This capacity to make adjustments, defined earlier as developmental acclimatization, is a part of the adaptability of the species and may be enhanced by the occurrence of human populations in areas where various forms of stress are repeatedly encountered.[3]

The adjustments are, in their aggregate, so successful that native high altitude dwellers can afford to be profligate in their consumption of oxygen, consuming more oxygen during exercise than lowland natives. It is possible that other, less understood factors are at work leading to this situation. One possibility is that the muscles of long-term residents at high altitude may possess more mitochondria, the organelles associated with oxidative metabolism. Increase of number and size of mitochondria has been observed in other mammals reared in hypoxic environments. The increase in mitochondria so observed is greatest in the so-called "red" muscle category associated with tonic rather than phasic contractions and may be important in the powerful action of the diaphragm at high altitude.

Observations of the growth sequence in the high altitude environment have shown that the growth of the trunk proceeds at a rapid pace during childhood, achieving nearly adult status as early or earlier than in lowland populations. Meanwhile, growth of the limbs lags farther and farther behind trunk growth in both sexes. There appears to be a gradient of growth within the trunk itself with the adult anterior-posterior chest diameter being attained early while the transverse chest diameter continues to grow. These factors, coupled with the lack of a pronounced pubertal growth spurt, result in a characteristic high altitude pattern marked by a high sitting-height index (another way of expressing the presence of short legs) and a large body mass/body surface ratio. The mass/surface ratio so obtained may be useful as a heat conservation device in addition to its other virtues at high altitude, since cold stress is also frequently encountered as well as hypoxia. While it is known that high concentrations of certain of the adrenocortical "stress hormones" delays growth and may lead to early closure of the epiphyseal growth centers, the role of these hormones in the determination of the growth pattern seen at high altitude has not been clearly established as a central one. However, there have been shown to be higher than normal (for sea level) concentrations of these hormones present in the adults of high altitude populations, a factor which seems to predispose many males of these populations to gastric ulcers. Retarded growth has been associated with a number of forms of stress known to induce high levels of adrenocortical hormone secretion. But in each case, other factors have been present to confound the situation, making it impossible to say that there is a clear-cut association between the level of hormone present and growth retardation, even though the evidence is strongly suggestive.

There are factors at work reducing the reproductive success of high altitude populations. But it is a general observation that native high altitude dwellers experience no unusual difficulty in achieving fertilization or implantation of the blastocyst in the uterine wall. There is an additional demand upon the placenta's gaseous diffusion function. Despite its status as a deciduous organ the placenta is a versatile and highly complex one. Not only must it transport oxygen and nutrients

3. For a general review of human adaptation to high altitude in the Andean area see: Paul Baker, "Human Adaptation to High Altitude," *Science* 163 (1969): 1149-1156.

to the fetus and remove the byproducts of fetal metabolism but it must also produce a number of hormones. While the ovary is the major source of estrogen, the placenta becomes the major estrogen producer during the third trimester of pregnancy. Some secretions normally produced by the anterior hypophysis are also produced by the placenta, including human chorionic gonadotrophin (HGT) which starts to appear as early as sixteen days after implantation. HGT secretion is reduced when the secretion of estrogens increases. A hormone very similar to luteinizing hormone (LH) is also produced by the placenta. LH stimulates continuous activity of the corpus luteum of the maternal ovary assuring continuous secretion of estrogen and progesterone preventing menstruation during pregnancy. Continuous secretion of progesterone causes a reduction in secretion of (LH) by the maternal hypophysis, thereby preventing an increase in secretion of follicle stimulating hormone (FSH) which could lead to an inopportune second ovulation.

Around the fiftieth to sixtieth day after fertilization, the placenta starts producing steroid hormones, releasing estrogens and progestins, as well as hormones very similar to adrenocortical stress hormones. It has been shown that these latter hormones may induce some resorption of maternal tissue to maintain the fetus under stressful circumstances. This prevents loss of a viable fetus as a result of malnutrition or other environmental stresses affecting the mother.[4] It has also been shown that the placenta has the capacity to rearrange its structure to enable it to attain a greater diffusing capacity when oxygen concentration in the blood is low. This feat is accomplished in part by ameboid movement of individual cells resulting in a thinner membrane with a greater surface area.[5] Also the small blood vessels of the fetus dilate, reducing the resistance to diffusion across the placenta and increasing the tendency for diffusion in the fetal direction.

Despite these capacities of the fetus to accommodate to lowered oxygen in maternal circulation, there are difficulties in supplying all of the fetus' demands at high altitude. This is particularly true toward the end of pregnancy when the capacity of the placenta to keep up with the fetal demand is taxed to the limit even in a favorable environment. As a result, there is a general tendency for babies born at high altitude to be smaller, on the average, than those born at sea level.[6] Infant mortalities everywhere are greater when birth weights fall below 2500 gm, a weight below which a newborn infant is termed "premature" even though it is not always certain that the gestation period has not attained the normal 280 day span. When such "premature" babies are born at high altitude, it appears that their chance of survival is better than those of similarly small babies born at sea level. Natural selection may be at work here, and we cannot reject the possibility that there may be a genetically determined tendency toward small babies at high altitude as a result. If this is indeed the case, high altitude populations may

4. W. G. Kinzey, "Hormonal Activity of the Rat Placenta in the Absence of Dietary Protein," *Endocrinology* 82 (1968): 266-270.

5. T. Tominoga and E. W. Page, "Accommodation of the Human Placenta to Hypoxia," *American Journal of Obstetrics and Gynecology* 94 (1966): 679-691.

6. Richard B. Mazess, "Neonatal Mortality and Altitude in Peru," *American Journal of Physical Anthropology* 23 (1965): 209-213.

have been allowed to survive long enough in their demanding environment through the employment of their capacity for developmental acclimatization to allow a suitable mutation or mutations to occur and be selected upon. So we may, in the high altitude environment, be witnessing a transition from an "acclimatized" to an "adapted" state. At this stage in the process and with our present sparse information concerning the genetic basis of a number of polygenic traits we are not in a position to say with certainty whether this is the case or not. Nevertheless, the populations living at high altitude provide us with an impressive example of the range of human adaptability in what is one of the most demanding ecological settings on this planet.

HUMAN POPULATIONS LIVING UNDER CONDITIONS OF NUTRITIONAL INADEQUACY

While hypoxic stress is encountered in a rather restricted area of our planet and the great plagues appear to be events of the past, the stress of malnutrition affects more people now than ever and appears to be on the rise throughout much of the world. Hunger is not new to human populations. Being relatively large mammals, we have a high metabolic demand. In addition, our evolutionary history has left us with a heritage of nutritional requirements which presents some qualitative as well as quantitative problems. As descendents of tropical, tree-dwelling primates, we have the expected need for the carbohydrates present in vegetable material and share with other primates the incapacity to synthesize our own vitamin C. In addition, some of our more recent ancestors acquired a taste for animal protein in amounts far in excess of that consumed by a primate since a remote, insectivorous prosimian form last wiped the severed legs of a cockroach from his receding chin. We, as a result, require the intake of certain dietary constituents in quantities not always easy to obtain.

Of the twenty amino acids regularly occurring in biological materials we must ingest eight. We are unable to either synthesize them or transfer amino groups and thereby produce them by conversion. Consequently, we have a list of eight "essential" amino acids which we must consume on a regular basis. Some of these amino acids are available in some vegetable foods. There are very few vegetable foods which possess them all, and those that do are not common. By judicious selection, it is possible to combine vegetarian foods in a manner which will provide all of the essential amino acids in sufficient quantity to avoid a nutritional imbalance. Few of the severely malnourished people in the world today have access to the selection of vegetarian foods found on the shelves of a natural food store and must make do on what they can glean from a sparse and overcrowded environment. In many respects, it is easier to supplement a diet deficient in one or more essential amino acids through the use of animal protein than by balancing vegetable nutrients, since animal protein, including meat, eggs, fish, and milk, contains essential amino acids in approximately the proportions most efficiently absorbed. Severe nutritional imbalances therefore are much more common in agricultural areas than among hunters and gatherers although starvation is no stranger to hunting populations in most parts of the world.

Some human populations subsist on diets that are almost entirely composed of animal protein, as is the case among Eskimos. The greatest hazard in such circumstances is vitamin deficiency. A number of devices have prevented severe vitamin deficiencies from decimating such populations. These include the consumption of intestinal contents of vegetarian sea mammals.[7] In most cases, Eskimos were in a well-nourished state before adopting a pattern of doing all or most of their hunting in supermarkets as has been the case among a number of them in recent years, to the detriment of their cardiovascular systems and dentitions.

In agricultural populations, particularly those living in tropical areas of the world, there is frequently heavy reliance upon some single crop to sustain life. This may be millet, maize, rice, or some other grain, plantains, or manioc. In good years these crops, while yielding sufficient calories, create imbalances associated with a deficiency of one or more essential amino acids and, in all likelihood, some vitamins as well. The most frequent vitamin deficiency involves water-soluble varieties of the B-complex, which function as coenzymes in a number of important biochemical reactions. In bad years imbalances are accompanied by inadequate quantities of calories as well, so that outright starvation may alternate with conditions of nutritional imbalance in a frequent and vicious cycle.

Our species has experienced feast and famine for much of its history. But imbalances may be a more recent experience and may therefore be less readily dealt with and may undermine the health of a population in a number of subtle ways. Shortage of sufficient protein reduces the capacity to produce antibodies, so malnourished children die of otherwise mild diseases. Vitamin deficiencies result in blindness and neurological damage.

There is evidence of a number of means by which humans may adjust physiologically to survive the stress of malnutrition. Adjustments may take place at a number of levels with responses falling into the categories of behavioral, physiological, and anatomical adjustments. We again see the presence of short-term physiological mechanisms as well as long-term physiological-anatomical changes giving evidence of a graded response system as a tactic of survival. Such graded response systems allow the organism to survive at the same time it retains a degree of flexibility to respond to other environmental stresses in addition to the one immediately at hand. Such mechanisms provide the essence of human adaptability to a shifting, frequently hostile environment. The ways in which humans survive severe malnutrition reveal several facets of the tactics evolved over time to deal with a threat to survival.

In adults there is considerable ability to buffer the effects of imbalance or starvation. Children, particularly younger ones, are more vulnerable to damage when malnourished but possess some short-term responses similar to those of adults. Among these responses is the utilization of protein from one's own body. There is a series of ways in which this may be accomplished. One of the first lines of defense in this tactical pattern involves the secretion of serum proteins, especially albumin, into the lumen of the intestine. Breakdown of

7. William S. Laughlin, "Genetical and Anthropological Characteristics of Arctic Populations," Paul T. Baker and Joseph S. Weiner, eds., *The Biology of Human Adaptability* (Oxford: Clarendon Press, 1966), pp. 469-495.

this rich source of amino acids facilitates the intestinal uptake of other amino acids present and, for a time at least, functions to correct deficiencies of essential amino acids. Serum albumin is soon exhausted in cases of severe malnutrition. This is one of the diagnostic signs in the onset of kwashiorkor in protein deficient children. When serum albumin can no longer be drawn upon, certain digestive enzymes, no longer functional in the absence of ingested protein, break down, releasing their constituent amino acids and for a relatively short time, correcting the imbalance.[8] Both of these defense mechanisms draw upon reserves which are much more ample in the healthy adult than in children. Reserves are most inadequate in infancy, a point of maximum vulnerability being during the post-weaning period. For this reason, the most frequent occurrence of protein-calorie malnutrition (PCM), seen in symptoms of kwashiorkor and other deficiency diseases is associated with the shift from a balanced diet of mother's milk to one of largely carbohydrate content.

In adults, prolonged starvation may be dealt with by drawing on other reserves including body fat, stored in adipose tissue and, finally, by the removal of amino acids from skeletal muscle. These amino acids may be used as such but also, in cases of starvation, may be converted to glucose by biochemical pathways of gluconeogenesis. As a result, even a fairly lean adult may survive for extended periods of low protein and calorie intake. Extensive studies of human starvation have shown that 50 percent or more of total body weight can be lost by a healthy, relatively lean adult without apparent permanent damage.[9] Of course, vitamin and mineral deficiencies complicate the problem of coping with inadequate protein and calorie intake and experimentation with the body's defense mechanisms by self-starvation can produce disasterous results.

One factor which makes the occurrence of starvation hazardous beyond the effects of inanition is that low protein intake makes continued antibody synthesis progressively more difficult. Antibodies are after all, proteins. As such they require amino acids including the essential ones in their structure. An assignment of priorities becomes necessary when a malnourished person is confronted with a severe infection. Should desperately needed protein be diverted to synthesize antibodies capable of dealing with the pathogen, or should continued supply of glucose and amino acids be favored? There appears to be a range of variability in both children and adults in the point at which antibody synthesis is reduced in favor of continued availability of nutrients. It is a common observation that the onset of the kwashiorkor syndrome is often associated with an infection and rise in antibody synthesis in malnourished children.

Malnutrition is a form of stress dealt with in many ways, and the list of responses includes the action of the adrenal cortex which produces increased quantities of antiinflammatory glucocorticoids. The adrenal response is in some respects

8. V. M. Salter, "Protein Metabolism," G. G. Duncan, ed., *Diseases of Metabolism* (Philadelphia: Saunders, 1965), pp. 9-14.

9. Perhaps still the most complete study on the effects of starvation is the World War II work described in: A. Keys, J. Brozek, A. Hanschel, D. Mickelsen, and H. L. Taylor, *The Biology of Human Starvation*, 2 vols. (Minneapolis: University of Minnesota Press, 1950).

antagonistic to the immune response. While these hormones reduce inflammation and thereby give symptomatic relief in many situations, their continued presence in high concentrations has the effect of reducing the production of antibodies. Viewed as a tactic, such reductions must be considered risky and, indeed, malnourished children frequently lose the gamble taken when priorities are assigned in this manner. In children, where growth creates demands more stringent than those experienced by adults, reductions in antibody synthesis generally occur more readily than in adults. As was mentioned in the discussion of high altitude stress, high antiinflammatory adrenocortical hormone levels are associated with reduced growth, presumably through reduction of protein synthesis.

Experiments with a number of mammalian species have shown that cortisol, one of the antiinflammatory hormones, causes a shift in liver function favoring the retention of amino acids by that organ and withholding of them from other tissues. Under normal circumstances skeletal muscle is constantly exchanging amino acids with the interstitial fluid. Amino acids from muscle eventually reach the liver, while others, which have passed through the liver, are brought to muscle tissue by the vascular system. Under the influence of high concentrations of cortisol, the liver retains amino acids while skeletal muscle continues to lose them. Since damage to the liver is potentially more damaging than that to skeletal muscle, it might then be considered that the muscles serve as a depot for amino acid reserves to be drawn upon under severe nutritional stress to enhance the individual's chances of survival. The overall effects of these adjustments differ in adults and in growing children, with adults having the greater storage capacity and lower demand for amino acids for anabolic (protein building) metabolism.

The greatest effect of inadequate protein intake is felt during the fetal and neonatal periods when some hyperplasia of skeletal muscle is still occurring. There is good reason to believe that inadequate amino acid intake during these periods, possibly up to the age of six months in a human infant, will be associated with a permanent reduction in the number of skeletal muscle fibers. When such reductions occur, they will usually be associated with a permanent reduction in the lean body mass (fat-free tissue mass) of the individuals involved. In the event of a later improvement in nutrition and an appropriate demand on the skeletal muscles through exercise, increase in the size of the remaining muscle fibers (hypertrophy) will allow the individual to perform work in a perfectly adequate manner. It is significant that when reductions in lean body mass occur, they are accompanied by reductions in both the protein and caloric dietary requirements of the individual.

In areas of endemic protein-calorie shortage, the individuals in the population capable of responding adequately early in life will stand a better chance to survive not only the critical stress facing them as infants but also long-term stress they will probably have to live with the rest of their lives. Individuals adjusting in this manner will be smaller than individuals not faced with the stress of malnutrition early in life. However, under normal work loads, which are characterized by activity levels considerably below maximal, these smaller individuals will experience little or no disadvantage. From the population's standpoint, the tactic of individual size reduction in the presence of endemic shortage of proteins and/or calories is distinctly advantageous since the overall demand of the population for scarce resources is reduced. Reduction in demand per person allows survival of a greater number of individuals on a limited resource base. These advantages are, of course, interconnected. The net result of the employment of this tactic is, through the maintenance of a greater number of individuals than might otherwise be possible, the retention of a greater amount of genetic variability through the minimization of natural selection. It appears that the overall effect of the exploitation of adaptability is the avoidance of adaptation.

As an example of the dimensions of the advantage gained by a population in this way, we see that, in a normal round of daily activities, a lean body mass reduction of 14 percent per male member of the population effects a 25 percent reduction in caloric requirements and a 14 percent reduction in protein requirements. Adult protein requirements appear to be capable of satisfaction by intermittent availability of high-grade protein. It is possible for an agricultural population through occasional slaughter of livestock and/or hunting and fishing to support one-third more individuals by the expedient of a 14 percent reduction in average lean body mass. If this means that 1,000 people can live where only 700 larger people could be supported, the populations stands an improved chance of being represented in the future gene pool of the species. The advantage might be lost if reductions in lean body mass were to impair the reproductive capacity of the population. There is reason to consider this to be a real danger.

The greatest impact of natural selection occurs at the level of reproductive success. Loss of fertility is the equivalent of genetic death. However, successful reproduction is itself meaningless unless the offspring themselves live to reproduce. Mammals are, as pointed out in chapter two, well equipped to insure that their offspring have the best possible chance to reach maturity. Humans have developed this mammalian characteristic to the greatest extent among living species. The capacity to maximize survival of conceptuses is facilitated by physiological characteristics of human mothers which protect the fetus and newborn during especially vulnerable stages of growth and development.

The embryo is especially vulnerable to damage. During the critical stages of cell differentiation and organogenesis, almost any disruption of the process will have far-reaching, probably fatal implications for the new individual. As the gestation period progresses, the demands of the fetus grow and, toward the end of gestation, may equal or exceed the capacity of the placenta to diffuse the substances involved across its membrane. Of course, these increasing demands ultimately must be satisfied by the mother. If the mother is unable to increase her nutritional intake, she must draw the necessary nutrients from her own tissues. Human mothers appear to be capable of maintaining a fetus even when poorly nourished themselves. It has been pointed out that the period of rapid weight increase occurring in human females before menarche adds a caloric reserve about 10 percent greater than the total needed to bring a fetus to term. There is evidence that the event of menarche is in some way associated with the attainment of this reserve.[10] The result of this association is that the human female is probably incapable of conception before she is capable of carrying the conceptus to a successful birth. Much of the reserve is in the form of fat, which can be broken down to supply the caloric needs of mother and infant as needed, cushioning the effects of inadequate nutrition during pregnancy. However, the fetus requires amino acids as well. Moreover, the demand on the mother does not end at parturition, rather it increases.

The period of nursing puts a greater strain on the mother's ability to supply protein and calories than does the period of gestation. A normal daily flow of milk (850 ml) requires the mother to supply 40 grams of protein and 1000 calories. The importance of this milk is critical not only from the standpoint of infant nutrition, but also for the formation of the infant's immune response. Mother's milk is a rich supply of maternal antibodies which allow the infant to cope with pathogenic organisms while establishing a primary immune response of its own. Where does a malnourished mother, who might well be ingesting less than forty grams of protein per day as well as insufficient calories, get the nutrients to produce sufficient mother's milk? Her only resource is, of course, her own tissues, tissues which are admirably equipped to release the nutrients needed by the infant under suitable hormone stimulation.

The system works well and undoubtedly is adaptive. It requires the accumulation, retention, and appropriate release of nutrients by maternal tissues. Skeletal muscle functions as a storage depot for amino acids and is therefore an integral part of a system for maximizing reproductive success under nutritional stress. It is therefore not desirable for females in malnourished populations to lose the storage capacity skeletal muscle fibers represent. So we see that in such populations, males exhibit reductions in lean body mass greater than that seen in females. The net result is a reduction in sexual dimorphism for body size which represents an effective tactic for reducing total caloric demand while maximizing reproductive potential.

Since growth tends to be allometric, reductions in lean body mass are accompanied by reduced growth in the skeleton as well. Therefore we see reduced statures in malnourished human populations but a general retention of body proportions which are probably the product of a long period of natural selection. Interestingly, relative sexual dimosphism for stature differs little in malnourished populations from that in well fed ones. As a result, women in areas characterized by endemic malnutrition frequently have a more robust appearance than their husbands.[11]

There are other aspects of adaptation to nutritional stress, including lowered activity levels. But the physiological and anatomical alterations already described

10. Rose E. Frisch, Roger Revelle, and Sole Cook, "Components of Weight at Menarche and the Initiation of the Adolescent Growth Spurt in Girls: Estimated Total Body Water, Lean Body Weight and Fat," *Human Biology* 45 (1973): 469-483.

11. For a detailed discussion of this and related phenomena, see: William A. Stini, "Adaptive Strategies of Human Populations under Nutritional Stress," F. E. Johnston, E. Watts, and G. Lasker, eds., *Biosocial Interrelations in Population Adaptation* (The Hague: Mouton, 1974).

are sufficient to demonstrate that human adaptability to ecological demands is many-faceted. If the discussion were extended to include responses to cold, hot, dry, sunny, cloudy, and crowded environments, examples of functional alterations involving a number of physiological systems would be possible. This is because we, as a successful species, are the possessors of a successful adaptive strategy which implies a suitable array of tactics for every ecological background regularly inhabited by the species. For discussion of human adjustment to a number of environmental conditions the reader is directed to the list of readings at the end of this chapter.

For Further Reading

Baker, Paul T. and Weiner, Joseph S. *The Biology of Human Adaptability.* Oxford: Clarendon Press, 1966. A collection of studies of the kinds of biological adjustments made by humans living in demanding environments. Useful for its range of physiological data.

Cohen, Yehudi A. *Man in Adaptation The Biosocial Background.* 2d ed. Chicago: Aldine, 1974. Provides the beginning student in anthropology with a collection of papers covering the range of adaptations from the strictly biological to the cultural.

Consolazio, C. Frank; Johnson, Robert E., and Pecora, Louis J. *Physiological Measurements of Metabolic Functions in Man.* New York: McGraw-Hill, 1963. A nuts and bolts approach to the function of the human organism under various conditions. A reference, certainly not a reader, but useful for the student interested in bioenergetics.

Edholm, O. G. *The Biology of Work.* New York: World University Library (McGraw-Hill), 1967. Consideration of a number of facets of human bioenergetics for the nonexpert. Provides some useful information about the day-to-day costs of human activity.

Harrison, Goeffrey A.; Weiner, Joseph S; Tanner, James M; and Barncot, Nigel A. *Human Biology.* New York: Oxford, 1964. Due for updating but still a useful exposition of human adaptability by a group of leading researchers in the biology of the species.

Bibliography

Appleman, Philip, ed. 1970. *Darwin.* New York: W. W. Norton Co.

Buettner-Janusch, John. 1966. *Origins of Man.* New York: John C. Wiley and Sons Co.

Comroe, Julius H. 1965. *Physiology of Respiration.* Chicago: Year Book Medical Publishers, Inc.

Hildebrandt, J. and Young, A. C. "Anatomy and Physics of Respiration." In *Physiology and Biophysics.* Edited by Theodore Ruch and Harry Patton. Philadelphia: Saunders.

Hughes, G. M. 1965. *Comparative Physiology of Vertebrate Respiration.* Cambridge: Harvard University Press.

Lowrey, G. H. 1973. *Growth and Development of Children.* 6th ed. Chicago: Year Book Medical Publishers, Inc.

Morris, Laura Newell, ed. 1971. *Human Populations, Genetic Variation and Evolution.* San Francisco: Chandler Publishing Co.

Osborne, R. H. and DeGeorge, F. V. 1959. *Genetic Basis of Morphological Variation.* Cambridge: Harvard University Press.

Stini, William A. 1974. "Adaptive Strategies of Human Populations under Nutritional Stress." In *Bio Social Interrelations in Population Adaptation,* edited by Johnston, F. E. and Watts, E. S. The Hague: World Anthropology Series, Mouton Publishers.

Van Liere, Edward J. and Stickney, J. Clifford. 1963. *Hypoxia.* Chicago: The University of Chicago Press.

5 How to Succeed in Adaptation Without Really Trying

How does one assess the degree of evolutionary success attained by a species? One frequently used criterion is the number of individuals representing the species present at a given time and a determination of whether this number constitutes an expansion, contraction, or equilibrium for population size. Expanding species and those seen to have attained equilibria at a larger number are thought to be successful. Related to the characteristic of population size is the matter of distribution. A wide range may normally be associated with success as a species. Obvious reasons for this latter relationship are the tendency for wider ranges to allow greater absolute numbers of the species to survive as well as the greater range of variability observed in wide-ranging species.

Man, by all of the above criteria, may be considered a successful species. The major threat to the future of the species appears to be environmental degradation resulting from overpopulation on a virtually worldwide scale. The success of the species is, of course, the result of successful adaptation to a wide range of environments. But the unity of the species has been maintained so that for 2 million years or more we may view human evolution as characterized by continued improvement of the species' adaptation. This improvement has occurred without the species splitting into separate lines. We may ask, "is the form of adaptation allowing evolutionary success for man unique?" to which the answer appears to be a qualified "yes."

In order to consider the human adaptive strategy most profitably it is necessary to begin with a careful consideration of what evolutionary success is. Large population size is not by itself success. Large numbers may, in fact, be a prelude to extinction. Rather, the most viable species are those which attain an optimum population size with respect to their environments. This means that consideration of the probable future of the species must be evaluated as well as the past and present. A species expanding at a rate which promises to exhaust essential elements of the environment or to contaminate its environment sufficiently to endanger future generations is headed toward a population crash and/or possible extinction.

The end result of such excessive growth is reduction in population size and a loss of genetic variability as the less adapted segments of the population are lost to intensified selection. Such losses of genetic variability reduce the overall adaptive potential of future generations. But, at the same time, they are an important evolutionary force since the process of specialization for any ecological niche must include "selection of the fit" members of the population.

But what if the strategy employed is to remain adapted to many niches or to a very broad niche? Or, viewed another way, how might a species most successfully adapt to an environment which frequently changes? Straightforward Darwinian selection, in which the population's gene pool is constantly reacting to circumstances prevailing one or more generations in the past, must incur heavy, recurrent loss of genetic variability in such situations. Adjustments which allow retention of variation will have the long-range effect of increasing the evolutionary capacity of the population while retarding the process of specialization to a given niche. Such adjustments permit accumulation of latent variability. They also allow a species to evolve at variable rates not usually predictable by deterministic models.

The possession of a series of nongenetic means of adjusting, including physiological, behavioral, social, and cultural mechanisms, has been recognized by many students of evolution as representing a feature most highly developed in, and in many ways, unique to man. But the difference between man and other species in this respect is more quantitative than qualitative. The extension of the human capacity to adjust into the realm of highly developed, technological cultures, while creating what, in many ways, is a new environment, with new selective pressures, has one feature that outweighs all others in importance. It increases survivorship. Despite the effects of war, famines, epidemics, and psychological stresses associated with contemporary life, the demographic trends in world population exhibit an inexorable increase in the number of humans occupying the earth. Viewed in this light, the adaptive strategy of our species, relying heavily on extraorganismic factors, has to be accounted a smashing success.

But more than just population numbers must be considered when long-term evaluation of a species' success is undertaken. More important is the maintenance of some sort of balance with the environment. This has, in man's case, come to include the entire biosphere. It is also essential to protect the genetic core of the system, even when environmental conditions change. There are those who view the attainment of control over environmental conditions of the sort characterizing contemporary human cultures as a sort of evolutionary dead end. Reasons for taking this position are, at first blush, compelling. Since man is so well buffered, the process of improvement of species' adaptation by well-defined Darwinian processes is, by and large, arrested. Superior genes may not necessarily displace their inferior alleles at many loci, since the environment no longer has the capacity to discriminate between alleles of slightly suboptimal fitness where selection is softened by overall improvement of the environment.

Reduction of selective pressure is usually accompanied by increased survival of marginal, suboptimal, and even semilethal genotypes. Cultural intervention has allowed survival of a much wider range of genotypes than have been hitherto salvable. Medical care, public health measures, improved nutrition, and reductions in death by trauma all tend to improve the survival potential of individuals who would, in other circumstances, be lost to the reproductive population of the species. In addition, demographic factors tend to reinforce the trend toward decreased selection pressure. For instance, in the United States in 1964, the average woman had a completed family of three children by the time she was twenty-seven years old. Sixty years earlier, it would have been more likely that a completed parity would include seven births with three survivors, with reproduction occupying the years between ages twenty-one and thirty-nine years. The tendency toward earlier onset and termination of reproductive life coupled with survival of almost all children born represents another area of relaxed selection. Infants born either very early or very late in their mother's reproductive span are susceptible to higher rates of neonatal mortality. The chromosomal aberrations characteristic of later reproductive life have little chance of providing useful genetic alterations which might in some way offset the losses incurred. The demographic shift toward earlier completion of families has resulted in a significant relaxation in selective pressure against neonates. Does this mean that human evolution has, for the most, ceased; or that the species is evolving only to better adapt to the cultural niche we presently occupy? Probably not. It is true that human adaptability has shifted emphasis away from straightforward Darwinian selection which would cause directional change toward specialization. But this strategy tends to increase variability rather than decrease it.

Our species has developed means of protecting its genetic core to a greater degree than any other species. The strategy employed involves social, cultural, and technological extensions of adaptive processes seen in other species. Most of the physiological and behavioral responses exhibited by humans exist in all mammals. Our species' adaptive strategy employs many molecular-level elements common to many forms of life. Protection of the genetic core is assigned a high priority in the allocation of resources. Development of a graded response sequence of increasingly more complex systems to accomplish this end follows a trend discernible throughout the history of life.[1] One by-product of elaboration of the system is the capacity to generate and retain heritable variation without paying a prohibitive price to natural selection. Species' success requires maintenance of species' variability if the demands of a changing environment are to be met. That this criterion has been eminently satisfied in man is becoming increasingly clear as improved techniques allow the identification of large numbers of polymorphic loci. It has been convincingly argued that when selection intensity at a locus is reduced to a negligible factor the number of alleles allowed is determined by the population size and the mutation rate.[2-5] Neutralization of many forms of selection and large population

1. For an interesting elaboration of this argument, see: L. B. Slobodkin "Toward a Predictive Theory of Evolution," R. C. Lewontin, ed., *Population Biology and Evolution* (Syracuse, New York: Syracuse University Press, 1968).

2. Matoo Kimura, "Evolutionary Rate at the Molecular Level," *Nature* 217 (1968): 624-626.

3. M. Kimura and J. F. Crow, "The Number of Alleles that Can Be Maintained in A Finite Population," *Genetics* 49 (1964): 725-738.

4. Matoo Kimura, "The Rate of Molecular Evolution Considered from the Standpoint of Population Genetics," *Proceedings of the National Academy of Science* 63 (1969): 1181-1188.

5. J. L. King and T. H. Jukes, "Non-Darwinian Evolution" *Science* 164 (1969): 788-798.

size combine in our species to make the presence of many alleles possible at many loci, increasing our pool of latent evolutionary potential.

Humans, being mammals and primates, have the advantage of being relatively unspecialized. Our species' adaptability has allowed the retention of this generalized condition even though the species has occupied several ecological niches over a period of time. We should be alerted to the fact that something different is happening in human evolution when we observe the ecological changes the species has undergone while exhibiting only the most modest morphological change.

The protection of variation has had the effect of storing up a large amount of evolutionary capital in the form of latent variability. As a consequence of reduced selection intensity, the species has reduced the pace of Darwinian evolution and allowed the accrual of potential for future evolution. The direction of future human evolution would, of course, depend upon the type of selection exerted by altered environments. Thus, the human capacity to alter the environment has more than passing interest for students of human evolution, particularly when we realize that we are necessarily viewing events in the role of participant observers.

For Further Reading

Crow, James F. and Kimura, Matoo *An Introduction to Population Genetics Theory.* New York: Harper & Row, 1970. Perhaps the final word, but difficult without a good mathematical background.

Ohno, Susumu *Evolution by Gene Duplication* Berlin: Springer-Verlag, 1970. A provocative exposition of ideas intrinsic to theories of non-Darwinian evolution. Rather difficult reading at times, but a stimulating piece of work.

Wallace, Bruce *Topics in Population Genetics.* New York: W. W. Norton, 1968. Clear and eminently readable compared to most books in the field. Probably the most accessible treatment of population genetics available.

Wright, Sewell *Evolution and the Genetics of Populations;* Vol. I, *Genetic and Biometric Foundations.* Chicago: University of Chicago Press, 1968. Tough going but will be consulted by evolutionists for the next generation and probably beyond.

Wright, Sewell *Evolution and the Genetics of Populations;* Vol. II, *The Theory of Gene Frequencies.* Chicago: University of Chicago Press, 1969. Not for the beginner, but a classic for those interested in the subtleties of evolution.

Bibliography

Dobzhansky, Th. 1970. *Genetics of the Evolutionary Process.* New York: Columbia University Press.

Dubos, Rene. 1965. *Man Adapting.* New Haven: Yale University Press.

King, J. L. and Jukes, T. H. 1969. "Non-Darwinian Evolution." *Science* 164:788-798.

Mayr, Ernst. 1963. *Animal Species and Evolution.* Cambridge: Belnap Press.

Glossary

Acclimation—A short-term functional adjustment to an environmental stressor.

Acclimatization—A nongenetic functional compensation to an altered environment as might be found in movement from one climatic zone to another.

Adaptive Radiation—The expansion of a species into new environmental zones with subsequent adaptation of populations in new areas resulting in increased intraspecific variation and, in time, the emergence of new species.

Allele—An alternative expression of a gene.

Altitude Thorax—A combination of anatomical characteristics seen in natives of high altitudes. Included are large intrathoracic volume, with barrel-shaped rib cage, long sternum, and associated increase in lung capacity.

Alveoli (Pulmonary)—The gaseous-exchange compartments of the lung.

Androgen—One of the hormones normally exerting a masculinizing effect.

Brachiation—Using the arms to swing from branch to branch in trees as a form of locomotion.

Carboxyhemoglobin—Hemoglobin combined with carbon dioxide to be transported away from metabolizing tissues and released in the alveoli.

Coenzyme—A substance, the presence of which is essential for successful catalytic action by an enzyme, as in the case of a number of vitamins.

Cooley's Anemia (Thalassemia major)—An anemic condition arising from a genetically transmitted defect in hemoglobin synthesis. Most common in Mediterranean populations.

Developmental Acclimatization—Physiological and anatomical adjustments to a stressful environment associated with growth and development taking place in the presence of persistent stress.

Differentiation—Production of specialized tissues and organs arising from cell division.

DNA (Deoxyribonucleic Acid)—A Polynucleotide sequence, or polymer, of a group of nucleic acids which, by alterations in the sequence of serialization, determines the genetic code of the individual

Dominance—A property of an allele causing it to express itself whether it is inherited from one or both of the parents.

Enzyme—A biological catalyst, facilitating reaction under conditions of moderate temperature and energy input.

Essential Amino Acid—The constituents of proteins which cannot be synthesized in sufficient quantity in the human body. These are: Arginine, Histidine, Isoleucine, leucine, lysine, methionine, phenylalanine, threonine, tryptophan, and valine.

Estrogen—A group of hormones produced, for the most part, by the ovaries, the presence of which has a feminizing influence on the developing organism.

Fitness—A term used in a relative sense to describe the influence of the possession of a trait on the probable reproductive success of its possessor.

Gamete—An egg or a sperm.

Genotype—The total genetic potential of an organism, the summation of its inherited characteristics disregarding whether such characteristics are expressed or not.

General Adaptation Syndrome—The combination of physiological adjustments, involving the secretion of products of the adrenal cortex, in response to stress-induced stimulation.

Genome—The total genetic potential of an individual or a population, but usually referring to the latter, where it it synonymous with the term "gene pool."

Genetic Load—The complement of potentially harmful genetic traits maintained in a population through various factors modifying or preventing selection which would eliminate such traits.

Habituation—Modification of response to a stimulus through repeated exposure.

Heterodont—A dentition characterized by the presence of teeth of varying shape in different parts of the mouth usually associated with specialized function.

Heterozygote—An individual inheriting a different allele of a gene from one parent than from the other.

Homeostasis—Maintenance of the internal environment of the organism within suitable limits through interacting physiological control systems.

Homozygote—An individual inheriting the same allele of a gene from both parents.

Hyperplasia—Growth through increase in cell number.

Hypertrophy—Growth through increase in cell size, usually used in connection with increase in the size of an organ.

Hypophysis (pituitary gland)—An important endocrine gland located at the base of the hypothalamic region of the brain which synthesizes or stores a number of hormones which have widespread physiological effects.

Hypoxia—Partial pressure of oxygen sufficiently low to induce stress.

Immune Response—Production of antibodies, proteins of the globulin variety, which act in several ways to combat pathogenic materials entering the body.

Intersexuality—Possession of characteristics of both sexes.

Kwashiorkor—A complex of symptoms generally associated with inadequate protein intake in children. Symptoms include edema, dispigmentation, loss of hair, skin lesions, lassitude, loss of appetite, and a number of serological abnormalities.

Lactobacilli—A group of bacteria which possess enzymes enabling them to break down milk to lactose.

Locus—A position in the nucleic acid sequence concerned with the determination of a single trait.

Menarche—Occurrence of the first menstrual period.

Mesozoic Era—The age of reptiles, including the Triassic, Jurassic and Cretacious Periods, from 225 to 70 million years ago.

Metabolism—The biochemical process essential to life. Refers to both "anabolic metabolism" involving energy storage and protein synthesis and "catabolic metabolism" which involves the release of energy through the breakdown of molecules.

Morphogenesis—The formation of anatomical characteristics by the process of cell division, differentiation, growth, and development, all under genetic control.

Mutation—An alteration in a genetically determined trait through change in DNA sequence.

Neonate—A newborn individual, up until ten days of age; infancy being the period from ten days to one year of age.

Omnivore—An organism capable of consuming a variety of foods to sustain life as opposed to those with rigorously defined dietary needs as in carnivores and herbivores.

Oxyhemoglobin—Hemoglobin to which is attached an oxygen molecule, usually the case when leaving the alveolar circulation.

Palate—The bony roof of the mouth separating the respiratory and digestive tracts.

Pathogen—An organism capable of producing a disease state.

Pentadactyly—The possession of five-digited extremities.

Phenotype—The result of interaction of genotype and environment producing the whole organism.

Phenylketonuria—A genetically transmitted, autosomal recessive trait characterized by a defect in the biochemical pathway for metabolizing phenylalanine. Buildup of phenylalanine in the blood causes brain damage and serious defect if not treated early in life.

Placenta—The deciduous organ, belonging to the embryo and fetus which exchanges materials with the maternal circulation.

Plagues—Epidemics of serious disease. Often used to describe recurrent epidemics which afflicted Europe during the period from the fourteenth through the eighteenth centuries.

Pleiotropy—Possession of more than one effect by a single gene.

Preadaption—Possession of a potentially adaptive trait before such trait confers any improvement in fitness. Useful in differentiating the Darwinian concept of evolutionary change which is preadaptive from concepts involving postadaptive change, implying inheritance of acquired traits, a generally rejected view of the evolutionary process.

Pronation—Crossing of the radius and ulna so that the palm of the hand is facing downward when the arm is extended outward from the body (abducted).

Range—The geographical extent of the territory habitable by a species.

Recessive—A gene which will not be expressed when paired with a dominant allele and is therefore only detectable in the phenotype of individuals inheriting it from both parents.

Recombination—The production of new chromosome combinations through random chromosome assortment occurring during meiotic cell division associated with gametogenesis in sexually-reproducing organisms.

Requisite variability—The variation present in a population allowing the action of natural selection in discriminating between more or less fit genotypes, thereby permitting evolutionary change to occur.

Sebacious glands—Oil producing glands found in the skin of mammals in association with hair or fur.

Secular Increase—An increase, as in average stature, occurring over time in a population.

Segregation—The principle that hereditary traits are determined by particulate, independent units.

Stereoscopic—A form of vision in which the eyes form images of an overlapping visual field allowing depth perception.

Stereotypic behavior—Behavior performed in a programmed sequence as frequently seen in birds and many lower animal forms.

Stress hormones—Several varieties of steroid hormone produced mainly by the adrenal cortex.

Supination—Positioning of the radius and ulna parallel to each other so that the palm of the hand faces upward when the arm is extended outward (abducted) from the body.

Uniformitarian—Change occurring according to normal, often imperceptible processes as in alterations of land forms due to the action of wind and water over extended periods of time.

Zygote—The fertilized egg which will divide to produce a new individual.

Index

acclimation, 10-11
acclimatization:
  developmental, 9
  high altitude, 56-63
acid-base adjustments, 55
adaptation. *See also* phenotypic frequencies of the peppered moth
  biological, 1, 6
  by adaptedness, 31, 53, 54
  definition of, 7
  evolutionary significance, 54
  extra genetic, 7
  genetic, 1, 4-7, 12, 36
  long-term, 6
  non-genetic, 2, 4
  preadaptation, 36
  reproductive success as measure of, 7
  shifting, 6
  "tactical," 6
adaptive radiation, 18
adrenal cortex, 8:
  adrenocortical response, 8
  androgenic steroids, 9
  keto steroids, 8
  secretions of, 8
albumin, serum, 65, 66
allele, 5, 73, 74
allelic variation, 2, 4
Allison, A. P., 42
allometric growth, 69
alpha chains, in hemoglobin, 39
altitude thorax, 61
alveoli, 54
amino acids, 15, 40, 46, 64-69
amphibians, 24
anabolic metabolism, 67
anaerobic metabolism, 58
anatomical adaptation, 29-30
Andes, 57-60, 63
Anopheles mosquito, 44
antagonism of adrenocortical and immune responses, 67
anterior neck muscles and human breathing, 55
anthropometric characteristics, 60
antibodies and disease, 48
aortic body, 55
arboreal:
  habitat, 27, 31
  human ancestors, 26

balanced polymorphism, 39
barometric pressure, changes with altitude, 53
barrel chest, 60
behavior, 11:
  adaptive, 11
  cooperative, 12
  learned, 11
  nonstereotypic, 11
  response, 11
  stereotypic, 11
beta chains, in hemoglobin, 39
biological investment, 8
bipedalism, 21, 31
birth, premature, 9
blood-buffer system and altitude adaptation, 57
blood chemistry, 10:
  acidity of, 10
  oxygen transport capacity, 10
  pH, 10, 55-58
  red blood cells, 37, 57, 59, 61
  viscosity, 59
body size:
  determination of, 4
  physical laws as limitations, 16-17
  and socioeconomic status, 3
brachiation, 28-29
brain, 23
breathing, mechanism in humans, 54
bubonic. *See* plague
buffered evolution, 72

caloric reserve in human females, 68-69
capillaries, alveolar, 55

carboxyhemoglobin, 37
cardiovascular system, 54
carnivores, 22
carotid body, 55
cell synthesis, 15
cellular metabolism, 15
central nervous system, 19-21, 27, 55-56:
  cerebral cortex, 21
cerebellum, and locomotion, 21
chicken pox, and American Indians, 48
cholera, 50
chorion, 23
classification, 15-16
clavicle, and brachiation, 29
contagious diseases, and sedentary populations, 47
Cooley's anemia, 46
core, genetic, 73
corpurs luteum, 63
cortisol, 67
culture, as a human adaptation, 31

Darwin, Charles, 34-35
Darwinian selection, 72
dead space, in respiratory tract, 56
dentition, 21:
  association with homeothermy, 22
  and dietary habits, 22
diabetes mellitus, 5-6
diaphragm:
  and human breathing, 54
  as a mammalian trait, 20
diffusing capacity of lungs, 59
digestive enzymes, breakdown of, 66
dimerization, of alpha and beta chains in hemoglobin
  synthesis, 39
disease organisms, 9

earth history, 34
ecological relationships, 6:
  of human populations, 44
  micro-differentiation, 18
  niches, 6, 12, 17-18, 72, 74
ecology, as an evolutionary factory, 6
embryo, 9:
  defective, 9
  retention of, 23, 68
embryonic period, 9
energy, 53
  transfer processes, 15
environment, 3:
  alterations, 4, 6-7, 9
  improvement of, 3
  stresses, 4, 8-9
environmental degradation, 71
enzymes, 15
epinephrine, and respiratory rate, 56
epiphyseal growth, 62
Escherichia coli, 35

esophagus, 20
essential amino acids, 64
estrogen, 63
eutherian mammals, 22
evolution:
  adaptation as, 52, 74
  "adaptive radiation," 18
  of bone and muscle, 17
  macroevolutionary, 6
  microevolutionary, 6
  overspecialization, 1, 6
  process of, 1
  success of, 6, 71-72
  variability, 6
extinction:
  examples of, 6
  probability of, 71
extragenetic factors, 4

famine, 4
fetal hemoglobin, 39
fetalization as an adaptation to altitude stress, 59
fetus, 9, 23, 69:
  nonviable, 9
  retention of, 23
fitness, phenotypic, 7
flexor muscles, and brachiation, 29
follicle stimulating hormone, 63
forced inspiration in humans, 55
forest clearance and malaria, 44-45

G-6PD deficiency, 47
gamete, 2
gamma chains, in hemoglobin, 39
Garn, Stanley, 4
gastric ulcers at high altitude, 62
genetic traits, 12
genetics:
  adaptation, 36
  "genetic load," 36
  science of, 34
  variability, 72
genome, 3
genotype, 2-4, 8-9, 12, 42-45, 72
genotypic potential, 10
gestation period, 9, 63, 68
glucocorticoids, 8
gluconeogenesis, 8, 66
glucose, 5
glutamic acid, 40
grazing animals, 22
group cooperation, 12
growth spurt, reductions under stress, 60
growth at high altitude, 62

habituation, 10:
  general, 11
  specific, 11

hand-eye coordination, importance to primate evolution, 27
heat conservation, 62
heme group, 37
heme synthesis, 46
hemoglobin, 36-37:
  A, 38-39
  alpha and beta chains, 39-41, 46
  C, 38-39
  equilibrium, 41-44
  fetal, 39
  molecule, 37
  S, 38-39
heterozygote, 2, 5, 38, 42-45
hibernation-reduced oxygen demand, 59
Himalayas, 43
homeostatic state, 10:
  control mechanisms, 53
homeothermy (temperature regulation), 19:
  association with activity level, 19
  association with dentition, 21-22
  association with increased range, 19
homozygotes, 5, 38, 42, 45:
  recessive, 5, 43
  thalassemia major, 46-47
hormones, 8-9:
  imbalances, 9
human chorionic gonadotrophin, 63
hyperplasia (cell proliferation), 9, 67
hypertrophy, 10, 59:
  muscular, 10, 67
hyperventilation, 10:
  at high altitude, 56
hypophysis, 9, 63
hypothalamus, 55
hypoxic stress, 38, 54, 59-62, 64

immune response, 8, 48:
  infant, 48, 69
  system, 47, 50
immune tolerance, 48
inanation, 66
infectious disease, 3
influenza, and South American Indians, 48
insectivore, 25-26
insulin, 5
intercostal muscles, 54
intestinal contents, consumption of by Eskimoes, 65
in utero environment, 9, 22
ions, excretion to maintain blood *pH*, 56
isoleucine, 46

Johnston, Francis, 5

Kelvin, Lord, 34
Klieber, Max, 53
Kunkle's fraction, 40
kwashiorkor, 88

lactic acid, 58
lactobacilli, 48
laminar flow, in arteries, 37
language, and human evolution, 31
latent variability, accumulation of, 72
lean body mass, 67
Lederberg, Joshua, 35-36
leucine, 46
life-associated elements, 14
limbs, positioning of in reptiles and mammals, 21
liver, 8
loci, genetic, 2, 4-5, 8
locomotion, 26, 28, 31
lumbar spine, and brachiation, 29-30
lung (alveoli), 20, 55-56, 59
luteinizing hormone, 63
lysine, 40

Macaque, 28
malaria, and sickle-cell anemia, 39
Malayan agricultural complex, 44
mammal:
  adaptive complex, 18
  dinosaurs compared, 18
  hair or fur, 19
  heterodont dentition, 22
  mammary glands, 19, 24
  marine, 18
  metabolism, 19
  muscular diaphragm, 20
  neurological integration, 25
  origin of, 18
  possession of placenta, 22-23
  reproduction, 22, 25
  sexual maturity, 24
  sweat glands, 19
  temperature regulation, 19
  tooth development, 21
  traits, 19, 25
marsupials, 22
maternal immune system, 23
measles, and American Indians, 47
  virus and mutual adaptation, 48
menarche, 4:
  occurrence of, 4, 69
  weight increase preceding, 68
Mendel, Gregor, 34
Mesozoic Era, 18
metabolic activity
  of green plants, 15
metabolic requirements, 36
mitochondria, 62
Monge, Carlos, 57
monotremes, 22
Morris, Desmond, 19
mother's milk, properties of, 69
muscles as a depot for amino acids, 67, 69

natural selection, 4-5, 68
neck muscles, and locomotory pattern, 30
neonatal mortality, 73
nonreproductive lifespan, extension of in humans, 53
nursing, 69
nutrition, 3:
  deficiencies, 9
  imbalances, 64-65
  infant, 69
  malnutrition, 65-66, 69
  requirements, 64, 68

omnivorous diets, 26:
  as a primate trait, 31
organism, 8:
  morphological characteristics, 9
  physiological characteristics, 9
overpopulation, 71
oxidative metabolism, 53
oxygen, 10, 53-57, 59, 61-62:
  atmospheric, 53
  "Bohr Shift," 55, 58
  deprivation, 10
oxygen consumption of high-altitude natives, 62
oxygen unloading by red blood cells, 55
oxyhemoglobin, 37, 57

parental care of young, 24
parity, averages in 1900 and 1964, 73
passive immunity, 48
pasteurella pestis. *See* plague
persistant acidosis, 58
phasic contraction, 62
phenotype, 2-4, 8, 12. *See* natural selection, 4
  altering, 4, 8-10
  frequencies in peppered moth, 35
physiological change, 8, 10. *See* stress response
  adjustment, 10-11
  physiological response, 10-11
PKU (phenylketonuria), 5-6
placenta, 22-23, 62-63, 68
placental system, 9
plague, 49-50
Plasmodium falciparum, 44
play, as a mammalian behavioral trait, 24:
  and learning, 24
pleiotropy, 3
pneumonia, 49
pneumonic. *See* plague.
pneumothorax, 54
point mutation, 41
polymorphism, 41-45
polyphosphate compounds, ATP (adenosine triphosphate), 15
polyribosomes, 47
population crash, 71
population size as a measure of evolutionary success, 71

preadaptation, 35
pregnancy at high altitude, 63
primary immune response, 48
primates, 25:
  and pentadactyly, 27
  Prosimii, 26-27
  traits, 31
progesterone, 63
pronation, and brachiation, 28
protein, 15:
  coenzymes, 15
  enzymes, 15
protein-calorie malnutrition, 66
pulmonary hypertension as an altitude adaptation, 59

quadriped, 30

range as a measure of evolutionary success, 71
red blood cells, and oxygen transport, 36
red marrow, and altitude stress, 56
red muscle, 62
replica plating, 35-36
reproduction at high altitude, 62-64
reptile, 18, 20, 22
resources, 7:
  competition for, 7
respiratory alkalosis, 56
respiratory rate, control of, 55-56
respiratory system, 54-55, 57-58
"Rhesus" incompatability (erythroblastosis fetalis), 23
right ventricular hypertrophy as an altitude adaptation, 59
RNA, defect in synthesis associated with thalassemia, 47
Romer, Alfred S., 18

scarlet fever, 49
sea level dwellers, 10, 58, 63
secondary immune response, 50
Selye, Hans, 8
septicemic. *See* plague
sex reversal, 9
sexual maturity, 24
sickle-cell anemia, 37-39, 46-47
shoulder joint, and brachiation, 29
smallpox, and American Indians, 48
specialization, 72
spinal column:
  curvatures in, 30
  and locomotion, 31
spleen, and hypoxic stress, 38
Srb, Adrian, 1
stature. *See also* hormones
  body size and, 3, 17, 69
  characteristics of high altitude natives, 60-61, 63
  determination of, 4
  secular increase in, 3
  sexual dimorphism, in, 69

sternum and brachiation, 29
stimuli:
  noxious, 11
  stressful, 10
strategy, genetic, 4, 6:
  as distinguished from tactics, 6-8
stress response, 8, 10:
  altitude stress, 61
  behaviorial level, 11
  G.A.S. (general adaptation syndrome), 8
  nutritional, 66-67
  "stress hormones," 62
supination, and brachiation, 28
survivorship, increase of in humans, 72
synthetic theory of evolution, 34

Tanner, James, 4
taxonomy, 15-16, 18:
  insectivore, 26
  primate, 26
  prosimin, 26
thalassemia, 46-47
thoracic diameter at high altitude, 60
thorax, and brachiation, 29:
  breathing, 54
tidal volume, 55
tonic contraction, 62
tool use, and human evolution, 31
toxic materials, 9
traits, 6
  maladaptive, 6
  mammalian, 25
transient polymorphis, 43
tree shrew, 26
typhus, 50

uniformitarianism, 34
unity of the human species, 71

vagus nerve, 55
valine, 40
variability, as an adaptation, 4
variable rates of evolution, 72
variation, as an evolutionary prerequisite, 2-3
vasoconstriction, 10-11
vertebrate, 16
  definition of, 17
  evolution, 18
  metabolic demands of, 17-18
viscosity of blood and altitude adapatation, 57

Waddington, Conrad, 4
Wallace, Bruce, 1
water-soluble vitamins, 65

zygote, 2